building
a
bigger
better
brain

NEURO-NASTICS

ROBERT MILTON Ph.D.

AuthorHouse™
1663 Liberty Drive
Bloomington, IN 47403
www.authorhouse.com
Phone: 1-800-839-8640

© 2011 by Robert Milton Ph.D. All rights reserved.

No part of this book may be reproduced, stored in a retrieval system, or transmitted by any means without the written permission of the author.

Published by AuthorHouse   03/01/2012

ISBN: 978-1-4567-5128-9 (sc)
ISBN: 978-1-4567-5127-2 (hc)
ISBN: 978-1-4567-5126-5 (e)

Library of Congress Control Number: 2011904312

Any people depicted in stock imagery provided by Thinkstock are models, and such images are being used for illustrative purposes only.
Certain stock imagery © Thinkstock.

This book is printed on acid-free paper.

Because of the dynamic nature of the Internet, any web addresses or links contained in this book may have changed since publication and may no longer be valid. The views expressed in this work are solely those of the author and do not necessarily reflect the views of the publisher, and the publisher hereby disclaims any responsibility for them.

Also by Robert Milton Ph.D.

TIPS—The Imaginative Parent Succeeds
Hole in the Soul (fiction)
The Unspoken (fiction)
The True Believers—the golden age of terrorism
The Reach (fiction)

DEDICATION

To my daughter
D'Elle and her husband Nick
Thank you for your unwavering support during
the most difficult five years of my life.

ACKNOWLEDGEMENTS

My last non-fiction book was published in 2005—that means this book has been more than five years in the making. During the past sixty months it has undergone innumerable changes not only because of personal issues but also because Genetic and Neuro-science will not stand still. Plus the insistence and inspiration of friends, family and professionals who simply would not let it 'be' as it was at several junctures. Still, I don't believe anyone ever catches up in the fields of Neuroscience or Genetics. In fact I'm sure of it. Too much new scientific information keeps pouring out of laboratories and universities all over the world every minute of every day—24-7.

I want to especially thank my bi-lingual computer engineer nephew (another USC grad) who uncovered some mighty small but very important inaccuracies and bloopers with his X-ray reading. As a Frenchman he would probably say *faux pas*. His expertise in several social and scientific fields made critiques, information correcting and gathering much swifter. I also offer my sincere thanks to Kay my delectable sounding board and comma coach extraordinaire. And especially my victims: Yes, the neighbors and other friends, who allowed me to read a very rough manuscript out loud in their presence, were of immeasurable help. Some of them even pretended to listen. (Of course, I read what was in my head rather than what was on the page)

And thanks to the artist Rivka who took a distant idea about brain plasticity and gave it "10 thousand words" as well as made it come alive with the kind of dreamy abstraction that would make Salvador smile. Then there is *Google*. One wonders why *references* or '*notes*' of any sort are really necessary in this day of immediate access to *Internet information*. But I suppose convention still plays a small part in the publishing business. So yes, I've included both.

IDIOTICON

Neuro-nastics:
Think REHAB for the brain. Like gymnastics, this term is used to describe repetitive activity and thought patterns specifically designed to literally modify certain brain structures. Athletes may do "carb loading"—we may all do rehab "content loading". i.e. provide new and/or repeated information for our brains—i.e. Neuro-nastics.

Nodule:
This term will be used as a synonym for brain plasticity. While it is recognized that *Nodule* is not a expression usually applied to any of the human brain lobes, anatomy or structures, still the word seems well suited to convey the meaning of *physically modified* neuro-anatomy as intended in this book.

Pollyanna:
Eleanor Porter's 1913 best selling novel describes a young girl with such an infectious optimism that it becomes a

contagion. Now considered a classic, Porter's title character's name has become a popular term for someone with the upbeat optimistic outlook, perhaps even innocently naive or of whom others could easily take advantage. Gradually the term Pollyanna has become an archetypal description of the tendency for people to agree with positive statements describing himself or herself or life in general. This bias toward what is called the '*positive*' is often described as the Pollyanna principle. Brain research has show that we are selectively attracted to '*optimistic ideas and events*' so much so that we tend to perceive the so-called '*positive*' even when it is not present in the objective world.

Uber-beliefs:

In this book *uber-beliefs* are considered as those thoughts, ideas, rites, repeated behaviors or rituals without immediate survival value, which become so pervasive in a culture so as to go unchallenged by logic or rational thought. Once an *uber-belief* is espoused, as an end-all answer, the search for a real-world solution is considered unnecessary.

GROSS ANATOMY of lateral view of HUMAN BRAIN

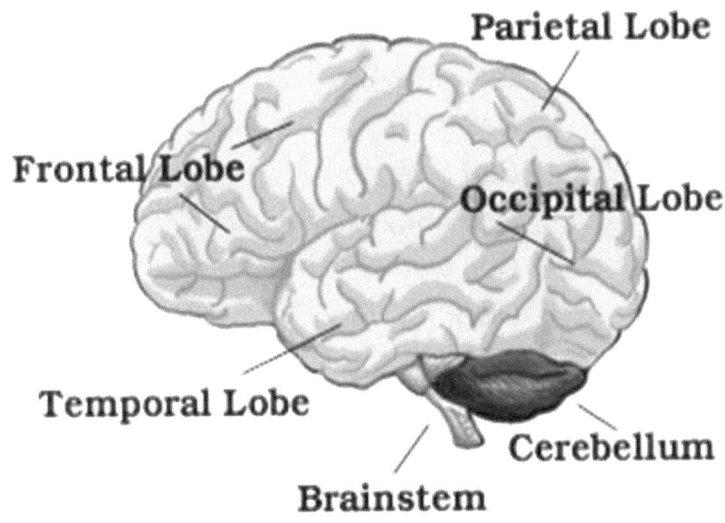

CROSS SECTION LATERAL VIEW

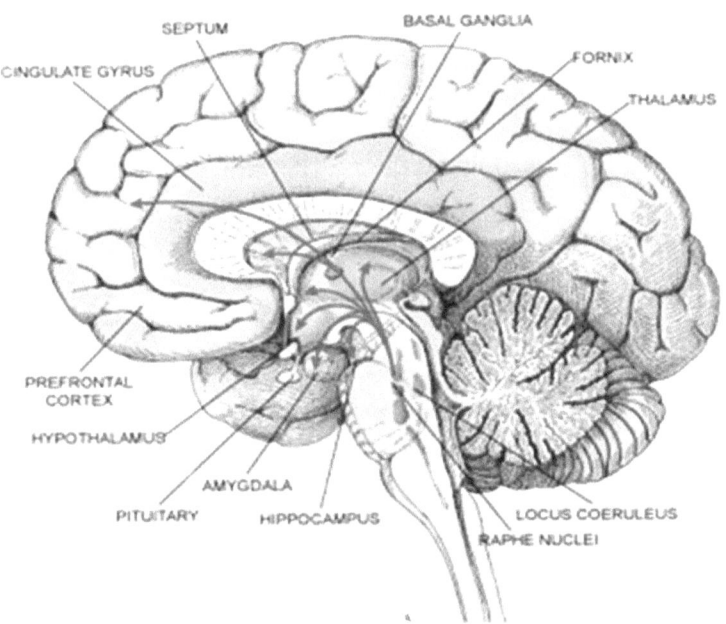

Robert Milton Ph.D.

<u>ACRONYMS</u> used in main body of text

ACORN	*A Completely Obsessive Real Nut*
BMA	*Blew Me Away*
BO	*Beige Out*
BS	*Bad Science*
CISTO	*Can I Say That Out loud?*
CO	*Cop Out*
DISTOL	*Did I Say That Out loud?*
GMAB	*Give Me A Break*
GS	*Good Science*
IDTS	*I Didn't Think So*
IGTP	*I Get The Point*
IMNSHO	*In My Not So Humble Opinion*
JJ	*Just Joking*
JMO	*Just My Opinion*
LOL	*Laugh Out Loud*
MHO	*My Humble Opinion*
MNBF	*My New Best Friend*
OMG	*Oh My Gawd*
PMIGBOM	*Put Mind In Gear Before Opening Mouth*
POV	*Point Of View*
SWEIN	*So What Else Is New?*
TIC	*Tongue In Cheek*
TISC	*This Is So Cool*
TFS	*Thanks For Sharing*
TMI	*Too Much Information*
TTTT	*These Things Take Time*
YKWIM	*You Know What I Mean*

Contents

ACKNOWLEDGEMENTS ix
IDIOTICON xi
ACRONYMS used in main body of text xiv

---
PART I   THEORY

OVERVIEW   FLEXING YOUR BRAIN 3
CHAPTER 1   OUR BRAIN **NODULES** 21
CHAPTER 2   THE POLLYANNA BRAIN 39
CHAPTER 3   FOR THE LOVE OF SILLINESS 61
CHAPTER 4   DUPED BY DESIGN 89
CHAPTER 5   WHEN IS NEW REALLY NEW? 109
CHAPTER 6   "RELIGIOUS—SCIENCE"
              & SCIENCE—SCIENCE 129

PART II   APPLICATIONS

CHAPTER 7   **NODULES** OF MODERATION 149
CHAPTER 8   THE **NODULES** OF INSPIRATION 167
CHAPTER 9   SURVIVING "AMERICANIZED"
              NODULES 185
CHAPTER 10   EDUCATING BRAIN **NODULES** 205
CHAPTER 11   "WAR" & OTHER HEROIC **NODULES** 231
CHAPTER 12   ADDICTION **NODULES** 245
AFTERWORD 261
END NOTES 279
RESOURCES & RELEVANT READINGS 359
INDEX 365

PART I

THEORY

The 1776 Revolutionary war was to set "us" free, the 1860 Civil war was to set "slaves" free, the 1920 Suffrage war was to set "women" free, the 1964 Civil Rights war was to set "races" free, and today a revolution is going on to free our "minds" from superfluous ideology . . . We can all participate and enjoy this revolution!

OVERVIEW

FLEXING YOUR BRAIN

Exercising means morphing

By now, it has been pretty well established that whenever a human *trait* or *characteristic* shows up on a worldwide scale among the majority of populations, we can be better than 90% sure that genes are involved. It is also probable that in the millisecond after human 'self-consciousness' appeared, questioning began. As if worldwide, we stood upright and immediately wanted to know and understand the great mysteries—understand the physical *'me'* and the spiritual *'it'*. Such questions appear to be worldwide in every age since history began. And let be clear from the beginning: An answered question, if it is authentic will stimulate another question.

Indeed, most human beings still crave understanding: Relationship understanding, "How the hell can I possibly get along with him, her or them?" We demand Political understanding: "What makes that SOB think he's a CEO or even a leader?" We pray for *Mystical u*nderstanding, "What's going to happen to me when I die?" We plead for *Religious* understanding: "Please <u>Reverend</u> tell me, what's going to happen to me when I die?" Right here at this modern historical juncture, a possible major human vulnerability raises its worrisome head. The word *Pollyanna* could be a beginning point to describe this. *Pollyanna* as used in this book, means our tendency to agree with positive statements when describing ourselves or countless other events. The phenomenon is related to research that indicates that at a subjective or subliminal level, our brains have a tendency to focus on optimistic material (probably related to survival itself); while at the conscious, objective level most of us delight in negative gossip (probably related to our elemental need to laugh)—at least for a moment or two.

In America our bias towards positivity is sometimes referred to as the Pollyanna politic. A close conceptual relative: *The Forer Effect* relates to the empirical observations that we most often give high accuracy ratings to descriptions of our own personality traits that were supposedly tailored specifically for us—whereas in fact, the traits were vague and general enough to apply to a wide range of people. This *Forer Effect* can provide a simplified explanation for the naive acceptance of beliefs and practices related to astrology, tarot cards, reading tea leafs and fortune telling in general. Yes, it seems many, if not most, of us are easily *duped*. We love mysteries and complex explanations that can be believed but never verified. Yes, I too can be included in this assembly.

It is certainly plausible that evoking something like a *Pollyanna* or possibly even a *"Dupe Gene"* complex within the hominid genetic repertoire would allow for our diverse understanding of the pain-full *human mystery* of how two rivals can gaze at one another over a border in numinous awe and still go to war. How is it that we can search for that sacred *something* in each other and ourselves—find it—yet still feel a self-righteous comfort when we deny the other what we ourselves most want?

SOME NEW ANSWERS

In the past, religious or philosophical mentors provided many of the more personal or private answers. They introduced us to *beliefs and uber-beliefs* to provide some calming and anxiety reduction. (Probably related to survival itself.) Then came psychological research. It offered empirically based answers related to anxiety reduction and human survival—mind-therapy. More recently most fresh, dramatic answers are coming from genetics and neuroscience. In verity, in contemporary times, neuro-physiology has become the *'sine quo non'* of new science.

Historically, Psychology had given itself to a philosophical cant that was supposed to bring an understanding about the human mind (i.e. *psyche-psychotherapy*). Then, when Psychology arrived on the American shore it quickly moved to an omnibus mode where everything and anything having to do with human behavior was considered fair game for Psychologists—as long as the research contained an "x" and "y" axis or fixed and experimental variables. TMI?

Lately, a broader based research technology is emerging called neuro-psychology. Neuro-psychology has filled in the intellectual menu where Theology and Philosophy were disenfranchised because they no longer satisfied the appetites of young '*questing*' academics yearning to 'think' rather than just 'believe'. As brain and body scanning instrumentation moved from the hospital to the laboratory, those young researchers who had been trained with an experimental bent quickly gravitated to what is now called the new empirical study of the *mind/brain*. They promptly came to understand that the human brain was not just a radio, (receiving signals from some other celestial mind or extraterrestrial dimension), but was the entire "Radio/TV Station" itself. The brain, they discovered, with its hundred trillion connections, was the *signal source*. These young scientists incorporated the new neuro-imaging (fMRI scans: *functional magnetic resonance imagery* and SPECT scans: *single photon emission computerized tomography*) to actually see pictures of how and where our hundred <u>billion</u> brain cells signal each other and how they interconnect with larger brain parts. They watched as they observed the human brain '*growing*' before their eyes!

Today neuroscientists can gaze at the molecules of emotion, as well as thoughts, as they dart from synapse to synapse (one hundred trillion brain connections) carrying out their never-ending *brain/mind* tasks. They watch as the unconscious becomes conscious. They now actually see the *mind* at its *multidimensional* work. One of the new findings, which we will talk about in detail, shows that we are not primarily products of conscious logical thought. The new brain view reveals the importance of *emotion* rather than *logic* in our interpretations of our world and a whole world of 'under the surface' machinery carrying out most human activities and thought.

Underlying these neuro-tasks are the new genomic codes that provide instructions for cellular and molecular activity. As a result, biology and psychology have undergone a genome-based revolution, which ultimately assists researchers to understand how slight, previously unnoticed changes at the cellular, molecular, genetic and bio-evolutionary levels produce extraordinary alterations at the behavioral-emotional level—although not usually seen as such. These advances are gradually changing our cognitive abilities just as at every moment in time we are improving our survival opportunities by continuously adapting to the events and discoveries of the day.

Because of recent discoveries related to brain plasticity we now have *Brain Spas* or *Mind Workout Gyms* where brain fitness exercises become a fun way to develop sharper thinking and extended memory.

Working-out and physical change is not just for changing the body; it applies to the brain as well since observable changes also occur in temperament and cognitive (*mind/brain*) processes. In this book it is called *neuro-nastics*.

You may wonder why it takes so long for this kind of new revolutionary information to reach most of us. First, today's neuro-scientific research findings are sometimes in conflict with more established academic information. When this happens it takes considerable time for the new data to be experimentally replicated dozens of times using the renowned but plodding *scientific method*. Second, more often than we care to think, the implications of neuro-research are in direct conflict with long held socio-cultural beliefs. As a result there is a general hesitation to study and publish it, lest we leave behind or offend the established *uber-beliefs*. Modern civilization's future success will not depend on science—that's easy. Altering *beliefs*—especially *uber-beliefs*—that's tough!

IMPLICATIONS OF *UBER-BELIEFS*

You've undoubtedly noticed that there is an increased need—or even a kind of panic—all over the world to understand the seemingly insurmountable discord existing between people, particularly where *belief ideology* is at the crux. However, when you think about it, the longest and most awkward conflicts have not been between different nations, or even different religions but rather, <u>within</u> the Muslim world and <u>within</u> the Christian world. Despite recent informed entreaties to interpret long-established scriptural writings by well-known ancient methods, the most caustic debates are typically between the ultra conservative Fundamentalists inside the major religions, who insist on a literal and static meaning of their *inspired scriptures* and the more liberal Modernists who generally want to scientifically legitimatize the *writings* or at least to interpret them metaphorically.

At one time or another, most of us have been told that *love* can transcend all, including these kinds of misunderstandings. Sure, that sounds clichéd but most of those who say it, sincerely mean it. For a case in point, I sincerely believe there is possible capacity in every human being to "bless the whole world—no exceptions!" But in the next breath we hear, "come on man, is that really possible in today's antagonistic climate?" By blessing '*them*' don't we give credibility to their system of belief and all the absurdities attending it? (See chapter 8)

Yes, understanding and appreciating *others point of view* can be really tough. And when we honestly make the attempt, we discover that most of the time *the big ME* (or what I later explain as brain *nodules*) is at the center of human problematic questing. Perhaps Leonard Cohen said it best: "*My friends are gone and my hair is gray and I ache in the places I used to play . . .*" In other words, what the hell is happening to ME? From the moment we realize our own mortality, we somehow feel a need to revisit conventional notions and *uber-beliefs* about another reality, a supernatural being, and an afterlife. What about us? Are we, or some ethereal part of us, immortal? In short, what should we do about the antagonism, not only between religions, but the heated hostilities between science and religion? At the center of *'it'* all is the big ME!

ATTEMPTS TO FIND ANSWERS

Many attempt to discover answers to *'it'* by expanding their consciousness via meditation. Then they find that classical meditation is a discipline requiring more time and self-discipline than they care to commit—at least at the present moment. Yes, you and I both know that's a CO (*cop-out*) for sure. So they seek a shortcut (as if such exists). A few chew a piece of cactus, or mushroom to bring about an alteration in the structure of their trillion brain cell connections. Such alterations result in what some perceive as greater sensitivity and a heightened perception of *reality*. However, the bona fide, transparent u*nderstanding* still remains elusive, which probably only shows our inward mistakenness. (Here I provocatively ask, "Does a fish thirst? Is this part of the cosmic joke?")

Recent neuroscience clearly shows—if we dare to look—these so-called *"higher"* experiences are already conditioned or determined by the growth or shrinking of brain structures that have been modified by what we have repeatedly allowed to enter our daily experience in our past; in other words, our previous brain '*content loading*'. In later chapters you will see that there is an increase or decrease in 'physical brain tissue' via *content loading.* This change is unmistakably revealed by fMRI brain scans.

As a short cut in this book, the word **nodule** is used to describe such changes in brain tissue composition. If our daily input and thought process is trifling, wary, and fretful, we may take the very latest designer mind-expanding drug or study the most profound literature but it is our own trifling, insignificant conditioned creation that is seen and understood. What is perceived as phenomenal or mystical insight can be shown by fMRI scans and the like, to be our modified brain structures (*nodules*) insisting that we *see* our own self-satisfying insights and epiphanies from our already conditioned history. At that moment we are *duped*. As a *'Pollyanna'* we may shrug and not bother to even wonder if we have a choice about this. Our so-called freewill appears to get swallowed up by our own conditioned brain modifications and genetic history. Understanding the ramifications of this becomes still another quest for a later chapter.

We may even boast that we want a 'special' sacred understanding that cannot be tailored or destroyed by progressive human research. So, once again behind this quest is our desire for the ultimate answers to—"What about *ME*? What is IT all about?" SWEIN? (*So what else is new?*) Maybe that's asking too much but it's the base question self-consciousness has always asked and still asks. At the same time, in this here and now, our very demand for an ultimate answer dictates what understanding may arrive. Talk about a tautology! For instance, I know I talk to myself but I am also aware that there is a conversation observer, or an *'other'* listening to my conversation—with myself. And most importantly, with the advent of brain scan technology—we can now look under the hood with motor running—we now know that *'other observer'* is <u>within</u> my hundred-trillion-neuro-connection brain. To suppose otherwise is to entertain the voices of psychosis or as a desperate dualistic alternative try to concoct elaborate metaphysical explanations, which, can be believed or disbelieved but never verified.

To have understanding, even a little, is for most of us a kind of transitory contentment. However, the more lasting, deep and wide our contentment, the more we experience delight. Bliss then, appears as a result of such understanding but then in a seamless heartbeat our bliss appears to dictate the form of our understanding. However, all too often we can have a certain kind of in-depth understanding as well as the ability to articulate it and yet be completely deluded, because inevitably we will only *see* according to our previous conditioning and brain structural modifications. (*Nodules*)

For example, we may *see* Christ, Buddha, mystical visions, wholesome ideologies, positive thought or whatever our neuro-conditioned belief systems promoted and still promote via repetition. The type of repetition is vital. Back in the thirties a brilliant homespun philosopher by the name of Ernest Holmes became the founder of a worldwide Religious Science movement by suggesting, among other things, "*Thoughts* were real and produce real consequences." Sure enough today it has been demonstrated that the more practiced we are in a "particular kind of thought" the more distinct and powerful are the changes in our own brain tissue. The external world becomes represented by these *changes.* Recent neurophysiology science demonstrates, our brains physically <u>grow</u> or <u>shrink</u> as we give or take away input—as we '*load content*'. Please allow me to again say: *Physical brain structures grow or shrink as you give or take away input.* For a case in point: REHAB takes 21-45 days to <u>begin</u> to effect a *nodule* change. It's not just a thought or abstract *mind* that changes—it is a 'flesh and blood' physical structural change! That's why *'content loading'* is so vital to our existence.

This same physical marvel (your brain) has been called the *harbor of consciousness*. This same two and half to three pounds of flesh and blood is where pictures of emotions and thinking, conscious or unconscious, can be taken and given over to empirical study. If human evolution has any validity, it follows that our consciousness is the consequence of basic molecular assemblies that evolved and can now be photographed and experimentally studied—just as any other physical science can be studied. When we eliminate the impossible, what ever remains, (however improbable in our well rehearsed belief systems) must be, at the very least, a closer proximity to this world's reality. What am I calling impossible? i.e. the whimsical nostrums—which can only be believed or disbelieved, rather than empirically validated and repeatedly observed by consensual validation. Myths and cultural folklore may comfort but more often they keep us uninformed by facilitating *uber-beliefs*. Sure, we may say we "make it all up" but when I *snip a key nerve within your noble brow* (prefrontal cortex) the *mechanism* that "makes it all up" stops "making it all up". So logically, if we would continue to evolve and make the unfathomable—fathomable and the unknowable—knowable and the incomprehensible—comprehensive, then we should want to know more about these basic *mechanisms.*

Our most pleasurable *understanding* then, is a bundle of tangible memory engrams—physical alterations, occurring in living neural tissue in response to stimuli—responding to our particular neural conditioning and can only respond according to our own physical brain structures, which we have literally created through repeated thoughts and actions. Furthermore, the cleverer we are at *loading our brain content* and then articulating our neural *understanding* the more our *understanding* reacts and breeds ideologies that are consistent with our own ever changing brain structures—*nodules.* The theories of Ernest Holmes and Donald O. Hebb (see Chapter 1) that supported this idea were apparently way ahead of their time. Today's recent discoveries of brain plasticity have been unparalleled because it now *'proves'* what Holmes and Hebb only suspected: Engrams are not just hypothetical. The human brain can be re-wired by *repetitive thought.* Or as is commonly said today: *"The neurons that fire together wire together."*

Here's a bit of irony: What a pleasure for me to find a book or TV program and/or read and quote some *'profound* guy' who agrees with my blissful understanding! (Facilitating my ever-growing *nodules*!) You may LOL, but if you get it, you've done it! So, in order for us not to get locked in or fixated in a particular personalized (no matter how comfortable and popular) neural loop or bias, we must continue to question, not only our *profound heroes*, but also our own ideological bliss. If we meet Buddha on the road, we know what we must do, because we all know that fish don't need to thirst. Don't we?

SHOCK AND CHANGE

My first university neurophysiology course was a shock—it changed my mind about a lot of things. The professor's opening line was "Mankind cannot fuck enough". Of course, he had the attention of every student in the room. He went on to describe the neurophysiologic consequences of orgasm. Again he had our complete attention! He used a few of the journal writings of St. Teresa of Avila, which he said were some of the best descriptions of orgasm in written record. Once again his statement allowed a few thought channels to fly open as never before. Tiny new *nano-nodules* undoubtedly began to sprout. Some others in the lecture hall, because of their previous neural *nodule c*onditioning, considered him sacrilegious. TTTT (*These Things Take Time*) In the Saint's journals she described a kind of movement of feeling through her inner "mansions" as a union with God. Indeed! Her memoirs have given theologians and pornographers much to think about. She described it as a "glorious bewilderment, a heavenly madness . . ." I would guess that Sis Teresa probably didn't know much Eastern ancient history, especially in the Arabian poetic literature, which is fairly oozing with such descriptions of "beloved God encounters." It also seems likely that such experiences would seem like utter madness to dedicated celibates.

Every sizeable *'faith'* in every part of this world has recorded experiences of a similar nature. Of course, they all use different metaphors. One could say that every great religion creates its own set of metaphors in order to convey its adherents to the presence of their particular Deity. Metaphors like poetry can indeed—transport. Most of the time serious Christian devotees called the derived transport "a visit from the Holy Spirit". Some Muslim poets refer to a "visit of the Beloved." Some Japanese mystics call it *ki*. In Chinese it is called *chi*. In the ancient Jewish Kabala tradition the encounter is described as a kind of spiritual ascension of energy up the spine along a series of invisible *chakras*. When I heard: "Just as there is in literature a literal fact and a poetic fact there is a physical anatomy and a poetic anatomy," it made sense. Many Yogis will tell you that these *chakras* do not exist in our literal body; they exist only in the meditative or metaphorical body. Still, some others seem to insist on literal, physical *chakra* spaces and will debate their actual physical dimensions. It seems that recent neurological research can help clear up this ostensible discrepancy. Meanwhile in the rest of world, there are many other words, phrases and metaphors used to describe the cerebral experience of neuro-movement into the pre-frontal cortex.

WHO IS IN CHARGE?

The *Overview* seems like the place to address a bit of urban mythology still existing in many cultures, especially those in which Priests, Gurus, Guides, Shamans and Spiritual Masters are idealized as being the surest and safest way to enlightenment. In much of India, for example, it is considered dangerous to venture into *blissful understanding* without supervision. The inexperienced, it is said, can literally wind up in a loony bin. The need for a guide and a safe place certainly seems valid and echoes the "60s turn on and drop out guru" (Timothy Leary) who instructed: "'*Set' and 'setting' are vital to tripping.*" And yes, seeking a teacher is certainly in vogue but it is also expensive and time-consuming. But *in my not so humble opinion* (IMNSHO) the really dangerous part is still being promulgated; namely that these particular experiences are the only true way to "encounters with some upward, outward divine supernatural nether-land dimension of something-or-other-unknowable-mystery". When this idea is disseminated as absolute, it can scare the bejeezus out of some, fixate others, and drive others stark raving bonkers. The enlightenment journey can be scary to some but as we realize via neuro-research, it is an inward journey to ones own previously created neuro-nirvana-divinity. That means that what has been allowed in *'there'*—your own experience and ideology—will manifest itself. Some of the most sincere guides, while *revealing* their ideas inadvertently concretize poetic metaphors making *them* an *'it'*. They ask you to repeatedly think *'it'*; the brain modifies *'it'* and voila! There you are—*'it'* is real and your new *nodules* insist you *believe 'it'*, repeat *'it'* and live or die by *'it'*. The *Pollyanna Brain* triumphs! *Nodules* grow accordingly.

CHAPTER 1

OUR BRAIN *NODULES*

Do we possess "free will"?

"50 million Frenchmen can't be wrong." Right? You've heard that aphorism and no doubt questioned its validity—with good reason. You know almost intuitively that *wrong* must always be compared to another point of view labeled *right*. Most human beings seem to innately possess a dualistic precept that *reality* consists of two basic opposing elements, often understood to be spirit and matter—mind and body—good and evil—positive and negative. In the case of Frenchmen we automatically read right and wrong. Why is *dualism* seemly so tied to innate human experience? Well, it has been pretty well established that with the advent of *self-consciousness*, we former troglodyte-cave-dwellers became aware that we could envision our own transience as well as our own inescapable death. In order to survive the acute mental pain (existential angst?) resulting from this frightening awareness, brain mechanisms were selected into our species genotype that helped reduce this distressing reality. A newly evolved neuro-mechanism (*The Pollyanna complex*) allowed—probably compelled—us to believe in another transcendent reality into which we *pass* or *transition* at death. In this alternate reality, we perceive ourselves as living on beyond our physical demise. That is, we live on forever in our choice of an idealized afterlife. Voila! Our '*demise anxiety*' is mollified and reduced and dualism is born! Consequently, we formulated a great variety of rites and rituals to help us remember this insight. (e.g. "The real you, is just passing through!") What is so surprising is that up until contemporary research and brain scan technology we had so misunderstood our brain and its function. In the past it was often called a radio-like receiver. Today we know the human brain is the whole radio/TV station with billions of transmitters. Also, until recently, our two and a half to three pound *harbor of consciousness* was considered pretty well fixed shortly after birth. Now we know that parts of it are physically changing all the time (yes, I said physically!) as we subject ourselves to various types of repeated '*content loading*'.

    The recent empirical discoveries of brain plasticity are unprecedented in human science because finally they now demonstrate that brain neurons could be rewired. Yet, from today's ongoing fMRI (*functional magnetic resonance imagery*) research we also now know that at any one moment in time, the average person uses not 10 but less than 4% of cerebral mass. Yes, it's true, at any one time for most mind feats, 96% of our brain's full collection of connective matter is not being used. This could be likened to having *The Presidential Air Force One* awaiting our willingness to learn how to pilot it. Instead, most of us choose to continue driving Model T Fords. (It was good enough for grandpa its good enough for me . . .) This may be because we "fix" ourselves into well-rehearsed *uber-beliefs* or familiar habits and thought patterns, which do not leave room for continued cognitive growth or even more importantly—brain evolution. OMG!

Let's return to the Messieurs of France. You also know that those 50 million Frenchmen have been nurtured by equivalent cultural/educational cuisine and there is a hefty likelihood that they may indeed be called *mistaken* when compared to another nation's mental cuisine. In fact, it seems reasonable to propose that once any person has been nurtured on a particular kind of cerebral food (e.g. *uber-beliefs* such as an idealized eternal afterlife into which we *pass*.) and has used those mental calories to repeat the informational exercises, (rituals such as prayers and meditations) that individual's brain undergoes physical changes that makes it all but impossible to ingest and process alternate kinds of informational cuisine. Mon Dieu! Once again nurture and nature collide and collude. Speaking of cuisine, it was recently discovered (Autumn 2010) that just a few weeks of food bingeing changes one's metabolism so that it is much harder to stay slim for years afterwards. Just a single month of pigging out on bad food, researchers say, makes for a lifetime fight with fat.[1] No wonder REHAB (*Neuro-nastics*) takes time!

EVER-CHANGING or EVER-GROWING?

In the late 90's newly employed London cab drivers were given a brain scan before their '*two-year crash course*' in learning and memorizing the street map of London—called *the knowledge*.[2] The drivers were again brain scanned after the map learning experience and it was discovered that an actual physical part of the brain known as the *hippocampus* had increased in volume. It plays an important role in encoding scenes (rather than faces or objects). Studies indicate that this region of the brain becomes highly active when human subjects view topographical scenes such as images of landscapes, cityscapes, or rooms—i.e. images of *places*. Damage to the area (e.g. trauma or stroke) often leads to a syndrome in which patients cannot visually recognize familiar scenes or topography.

In just two years of mental indoctrination (Neuro-nastics) the brain scans of taxi of London cab drivers showed amazing changes. These changes appeared to confirm the previously discredited view of Canadian neuroscientist Donald Hebb[3] who in the early 50's prophetically suggested that brain circuitry could be modified and repositioned so that literal brain structures could be physically changed. As mentioned D. Hebb and E. Holmes were way ahead of their time. Other brain researchers such as Lashley, allegedly laughed at Hebb. But then, as we say, hindsight is 20/20 and Hebb's supporters have had the last laugh.

HEART OF THE MATTER:

When I read the decade-old research about London cabbies, I asked myself a question: If human beings are subjected to certain kinds of education, indoctrination, and programming and the structures of their brains physically change (in this case enlarge) to accommodate such learning, **would the physical modification in their brain structures hinder alternative ways of seeing, understanding, experiencing and questioning the events of their current world?** In other words, once human brain structures have been physically modified, the question remains: Do such modified brains in these individuals have what is popularly called *free choice* or *free will*? Answering this question becomes the heart of the book you are now reading!

A classic example might be, observing a Muslim or Catholic child subjected to a routine catechism of religious doctrine and required to perform rote-memorization of ancient scriptures for several years. According to recent neurological research that child's <u>physical brain structures</u> change as a result of such religious/educational programming. And an important question remains: Does that youngster at the age of twelve or twenty have the brain capacity or mental aptitude to make choices other than those consistent with what he or she has been indoctrinated to believe?

Furthermore, if peers, parents, mentors and ministers produce and deliver that same-programmed spiel, it seems very unlikely that such an individual would be <u>able</u> to see another view of the world or seriously entertain other religious points of view. Such individuals may verbalize a sincere desire to explore and understand other points of view but those same individuals would not or perhaps <u>could not</u> readily make a choice to do so. New brain structures (*nodules*) would have developed that provide those individuals and "fifty million Frenchmen" a particular (peculiar by some standards) way of viewing the world and their own futures.

In this book I will repeatedly use the word *nodule*[4] to mean any physical brain structure modification. Dr. Hebb, mentioned above, used the phrases *cell assembly* and *phase sequence* to cover similar modifications but since such alterations could not be photographed at that point in history (1950s), his idea was discredited. Now it appears that *nodules* produced as the result of indoctrination and acculturation make the possibility of re-considering his thesis a point-at-able reality. The *scanned brain 'nodule'* (assembly) can now be seen. When years of indoctrination and programming are added to the mix, it is little wonder that for some, considering other points of view becomes effectively impossible. Or if possible, it would be achievable only after some months of intense alternative study. (*Neuro-nastics*) For example, it is pretty well accepted that REHAB programs are rarely successful when practiced less than two years.[5] GS. (*Good Science*)

In recent years, neuropsychological research has progressively and repeatedly demonstrated that environmental conditions, education, and intense study do indeed shape brain tissue structure and functioning. (i.e. *nodules*) In their 2001 classic college textbook Bloom, and Lazerson[6] say, "Experience [learning] can cause physical modifications in the brain." Still more recently, this BMA (*blew me away*), German researchers[7] found that simple three ball juggling increases the tangible size of the human brain. Not only physical acts like juggling can create new *nodules,* but also common, almost ubiquitous, mental exercises can do the same. For example, learning a new language can do it and the lack of mental working-out can actually reduce human brain structures.[8] Again the brain is indifferent as to the type or quality of *content loading* we perform. Back in the thirties, Ernest Holmes articulated a similar thought in that it doesn't matter what the content is . . . the mind will *act on whatever we suggest.*

Today neuroscientists like M. Stryker of the W.M. Keck Foundation for Integrative Neuroscience are saying things like: "The brain prioritizes data [without passing judgment. It learns what is there to learn]. Parts are in constant competition to see which ones can create lasting changes. Where you focus your attention helps determine where positive plasticity (*nodules*) takes hold."

Andrew Newberg's impressive research, condensed in a book entitled *Why God Won't Go Away*[9] provided insight into the workings of the brain when it is deliberately being subjected to an intense '*meditational workout*'. (Another kind of *neuro-nastic*) Both Catholic Nuns and Buddhist Priests were examined via a SPECT scan (Single Photon Emission Computerized Tomography) during a peak moment of meditation. Newberg and his associates found some dramatic changes taking place in the left side of the brain just below the crown and slightly toward the back part of the upper cortex—usually called the left parietal lobe. He called it the OA or orientation/association area because research has shown that this area is essential to keeping ourselves upright, walking up and down stairs, juggling balls, and orienting our bodies in life's common daily encounters like hugging or shaking hands.

The SPECT scan showed nourishment leaving the OA area and flowing into the pre-frontal lobes where imagination, abstract thinking can work their 'magic'. Remember that without OA nourishment the individual would feel like there were no boundaries. Since the pre-frontal area is responsible for big chunks of forward thinking, fantasy and awe the individual would attempt to account for a boundary-less existence. Future planning, fantasy, creativity and imagination are really hard to untangle. It's only a tiny step to appreciate and understand why—individuals who allow their brain to move into one of these alleged mystical *one-with-all-transcendent* experiences—at that brain *enhanced* moment they truly believe they are *one* with all-that-is-in the presence of the *'One'*. The pre-frontal cortex is the part of the brain where novel, innovative imagination can run wild! In their minds they are only a micro-step away from realizing intimate contact with their indoctrinated concept of immortal gods. During these mind-altering practices, which are by rational standards pure products of their own brain structures, they *feel* no boundaries—it is an ineffable experience.

But, wait a minute . . . for radical *believing* Muslims as well as some fundamental Christians this kind of research and understanding showing the *'contact'* is *inside* the brain rather than *outside* is probably sacrilegious. Or, as one neuro-scientist said to me last summer, "Denial of *outside contact* doesn't sell books." For most "True Believers" dissent would set in and they would vehemently repudiate any such possibility. Once again, in the context of indoctrination and life-long mental exercise (*neuro-nastics*), more likely than not, their well oiled *beliefs would* join with their highly developed *nodules* and in the end *uber-beliefs* would triumph.

These changes in brain nourishment during meditation compel the individual to feel a lessening of the borders between themselves and the physical world. In fact, they repeatedly report feeling a loss of boundaries and a physical blending with all of reality. In short, they reach a state of ineffable, emotional *Nirvana* and *oneness* with all. It is little wonder that some individuals having this experience describe it as "meeting the *'One'* face to face". OMG!

Such heightened emotional experiences do not appear to be beyond the range of anyone's meditative brain function. However, discipline and years of practice are usually required. A few others have seemingly arrived at similar states by the use of hallucinogens such as Peyote or LSD. These short cuts are taken, they say, because it's like a person on a quest to reach a mountain peak. A fog bank envelops the whole mountain. The person becomes discouraged and considers giving up altogether. Suddenly there is a break in the fog (the psychedelic experience) the peak appears in all its glorious luminescence. The person takes heart and continues the quest. Individuals ingesting hallucinogens repeatedly report that they have met with God. Chemical ingestion leaves little doubt that such meetings are indeed "mental" experiences.

Some critics have laughingly suggested that the insipid forms of prayer and meditation practiced in the

West would not impinge, on what Newberg called the AO, in the slightest. In the Muslim and Buddhist world however, it may be that a twelve-year-old, practicing repeated religious thought patterns several times a day could produce a dramatic SPECT image.

No doubt a brain scan of a dyed-in-the-wool radical terrorist would reveal that his so-called divine directives to kill infidels were a result of his mumbled prayers and incessant chanting. Could he, would he, see this possibly? Would this book make any sense to him? Because of his ever growing *nodules,* could he understand that his inspiration is neurologically derived and has no connection whatever with an otherworld Allah *speaking from paradise* or the supernatural?

WHY WE ARE SO FIXED IN OUR BELIEFS

Once we *see* and begin to understand more fully the miracle of our brain—the *harbor of consciousness*—and appreciate that an estimated 96% of our brains mass is going unexploited at any one moment, we can begin to see how provocative and engaging neuro-science really is! In this day and age it is ill conceived to believe that *ignorance is bliss* or that *magic* or Magicians provide real answers. We have a Presidential 747 at our disposal—why keep driving a Model T?

Perhaps our *free will* in this matter has been impacted by our brains interpretation of available data.

What data? Your genetics, your history, your social and cultural situation, your memories, your personality and how other people respond to you and particularly your daily *'brain content loading'* plus a myriad of other factors. All of these and more have been repeatedly impinging on your life—and most particularly upon *your brain*, which by colluding with your thoughts and life style is constantly modifying itself. By creating comfy, subjective neural habits, we think and behave in a seamless fashion. What, in this book is called *'nodules'* form new brain features and once again, while we may assume we are *free to choose* another path or point of view, we now know that we exist *within* many dynamic controlling variables. It begins to look less and less like *free will or choice is ours to exercise!* After two or three years of repeated thought we are probably stuck. Unless . . .

BRAIN AND BODY COMPARED

The structure of a human body, as we all know, is mostly determined by genetics, nutrition and physical exercise. For example, years of lifting weights allowed Arnold Schwarzenegger's muscles to change dramatically. A person, whose physical activity centers on accounting or word processing, will in all probability not have Arnold's muscles. JMO, *(just my opinion)*. In much the same way, recent neurological research suggests that the *type and amount* of mental exercise a human receives, determines the location, size and shape of their brain *nodules.*

At this point we might argue that little Mary Lou would never look like Arnold Schwarzenegger, even if she spent twice as much time in the weight room. It is true that the basic blueprint of Mary Lou's body structure has been determined at her conception by genetics. No one would deny that, but it is also true that if Mary Lou spent the same amount of time with weights, nutrition and hormones as Arnold, her body would certainly look different after a couple of intense years of concentrated exercise in the gym. *Neuro-nastics* and gym-nastics have a lot in common.

Humans are all born with bodies that look somewhat different. In the same way, humans are born with neurological differences. We all have different aptitudes and potentials. Yes, it can be harder for some kids to learn to read or do math and there will be those who can read or do math very early, and easily as well. Obviously, we cannot all be soccer stars. In fact, some find it hard to learn to play skillful soccer. But by following a *correct method of instruction* and with *sustained ongoing practice*, almost any person can learn to play basic soccer. In the same way most people can learn to read acceptably if the correct method of instruction is followed and if enough practice is provided.[10] But, it takes a whole lot of practice plus a concerted willingness to consider change. Even if change is desired, extended REHAB (Neuro-nastics) can be tough!

WHY "CHANGE-OF-MIND" IS SO DIFFICULT

In another book[11] I outlined a few reasons why we as human beings are so hesitant to change. The first was the need for "sameness" or status quo. When Bruno was burned alive and Galileo was imprisoned for life because they announced their scientific discovery: *"The sun is the center of our solar system."* Those in power merely wanted to maintain the current status quo regarding their understanding of the religious/scientific dogma of the day including the solar system. They were trying, at almost any cost, to fulfill the rather basic human need. The political and religious leaders of Galileo's day were no more committed to do evil than are comparable leaders of today. The drive for sameness or status quo is a powerful force in every generation. "Better to dance with the devil we know . . ."

Second, was the human need for authoritative instruction and/or commands. Obviously a majority of human beings are *herd animals* and need to *believe* that information gleaned and passed on over centuries of time, distorted or not, has veracity and is applicable to contemporary civilization. It is amazing how we trust almost any self-proclaimed authority to interpret what we consider to be mysterious, especially if they show us some "magical qualities". (Probably the *Pollyanna complex* in action.)

Why, you may ask, do we so passionately believe that which is so palpably erroneous?[12] I suggest and will elaborate further in chapter 2 that certain basics, instilled at a very early age, have imprisoned us. You and I and every child on earth depend on those basics. In part, they are the *uber-beliefs* some are willing to die for. Almost every person on earth has been, at one time or another, culturally indoctrinated to believe that a 'special' someone in the role of leadership knows the *secret secrets*—has the answers. Here, those who really *'know'* may begin to LOL.

It may also be that we are hard wired to resist change. A researcher at the University of California at Los Angeles has suggested that *Change is Pain.* Professor Jeff Schwartz and others have shown that when what we *believe* will occur and what actually happens, are different, in *Pollyanna fashion,* it is reported as *error by the prefrontal cortex* in a specific area that appears directly linked to human fear and anxiety. Our ability for higher reasoning becomes abridged and we desire above almost anything to retreat into familiar territory. This appears to represent a natural, evolutionary tendency to resist change and seek a familiar "comfort zone" of well-worn beliefs.

As suggested in the above neurological research, *changing one's mind*, that is, changing fully developed brain *nodules* is probably <u>not</u> a matter of immediate choice. This is where *Neuro-nastics come in.* In this country we have created hundreds of REHAB facilities because we know that without weeks and months of intense rehabilitation and re-learning the addict has almost no chance at all. Perhaps it could be proposed that the same is most likely true for most of the conflicting ideas in the world especially the Mid-East Theocracies verses the Western Democracies. We may think, if only they could be rational, if only they wouldn't be so stubborn etc. etc., we wouldn't go to war with them. Which of our leaders has suggested that not only do they not see our point of view—they cannot see it because their brain structure *nodules* prevent them from seeing or understanding any other point of view than the one they have been trained and indoctrinated to see? Is it possible that cultural "radio stations" have been allowed only one script and have been repeatedly broadcasting it without pause for generations? Can you say it is any different in Western countries?

After the longest war in US history we should certainly know by now that even extreme 'arm-twisting' does not work where long-standing religious beliefs (*nodules*) are concerned. Contained within our new pursuit for peace there needs to be an attitude of sharing, educating and making the new 'radio-broadcast scripts' available in a palatable manner. How? Well, how did we all come to see that the Sun, not the Earth, is the center of our solar system? That scientific insight certainly went against the entrenched beliefs of the day.

Perhaps it may not be easy to see the answers clearly and immediately because today we earth creatures appear to be swimming in a sea of giant sized *dupers* and *opportunistic predators* using every medium available (including science, politics and religion) to compete for the chance to trick, threaten, titillate, terrorize and—as if possible in today's polluted waters—to *insult us*. Getting even one partially submerged insight into these absurd human antics can buoy us in a bubble of laughter. YKWIM? *Neuro-nastic* training in the subjects of *doubt* and *curiosity* can be a beginning.

CHAPTER 2

THE *POLLYANNA* BRAIN

Why we love magic and illusion

Musicology, that is, the theory of music was one of the areas where I, as a new college student, felt an absolute blank in my life. My *tin ear* made it pretty obvious that the *music gene* didn't take a seat in my chromosomes. When I discovered I had a choice about certain elective courses I thought it would be fun to learn a little music to fill in one of the voids. The professor was a genius on the organ and piano. After listening to him play Cantata 140: *'Sleepers wake, the voice is calling'*, I thought, "This is magic! Yes, I need to learn more about this kind of music, even without a chorus this Bach guy is really mystical."

I sat in stunned stupid silence for the first couple of lectures and realized this particular 'magic' was way beyond anything I had ever encountered. The Prof had a fanatical affection for Bach that in my mind surpassed all reasonableness. I barely managed to hang on and then, just when I thought I was going down for the last time, the third lecture began and that brilliant professor deconstructed one of Bach's cantatas and visually demonstrated to the class the various underlying structures of the music and, lo and behold, I could actually see the essential mathematical arrangement of that cantata and suddenly it became transparent. Clearly, Bach was a mathematical genius and nimble of finger but now, at least to me, this wonderful bit of music was revealed. How to capture this bit of divine-like inspiration? I know that neither Bach nor his music was paranormal—definitely not a mystical illusion either—IMHO. (*In my humble opinion*)

Most of us have met people who do not wish to probe some matters for fear that they will indeed *lose the magic*. In fact, a good case can be made demonstrating that many human beings have a gene complex that predisposes them to love the mystery and trickery of magic. Some find Las Vegas illusionists irresistible; magicians and hucksters receive their adulation and money for deluding them. I am often among them. Furthermore, some of my fellow-*Pollyanna's* don't want magic tricks explained because they believe that explanations detract from the *awe* and *wonder* of it all.

Richard Dawkins, in his stimulating book *Unweaving the Rainbow*,[1] addresses this so-called *detraction* and the misperception that science and art are at odds.

Science, it was said, deprived life of deeper artistic meaning. Apparently, Dawkins felt a need to explain that misconception. As a scientist, he sees the world as full of wonderment and surprise as well as a source of pleasure. This pleasure is not in spite of, but rather because he does not assume *cause* is inexplicable but rather contained within understandable laws of nature. Keats's well-known accusation that "Newton destroyed the beauty of the rainbow" becomes the book's beginning point, then Dawkins' shows the reader that science does not destroy, but rather discovers the poetic and further artistic inspirations discernable in the awe-inspiring patterns of nature. Allegedly Dawkins said, "I am against religion because it teaches us to be satisfied with not understanding our world." I ask, "Could this possibly be true?"

GENOMICS

Recently, especially since the *cracking of the genome*, much has been made of genetic possibilities including a solution for the nature vs. nurture[2] debate. Because it's happening so rapidly, it is difficult for many of us to grasp the changes that have already taken place in various scientific fields. While we may grasp a tiny bit when they tell us, "all the genetic material in the sperm and egg cells that produced the Earth's present population could fit into a space the size of an aspirin".

WOW! Perhaps by such comparisons we can also grasp a little understanding and have a vague idea of what's happening in genetics. Because popular journalism makes such dramatic statements our attention may be directed to some exaggerated new notion. However, in spite of countless questions and some *unknowing*, we had better get ready for momentous changes because genetics has been wed to neuro-science. What does that mean to you and me?

NO GENE IS AN ISLAND

Recently, notwithstanding the general knowledge that no single gene acts alone, a few reputable researchers have given names to various genes including the *God gene, Criminal gene* and even a *Gay gene*.[3] So, is it possible there may also be a *Dupe gene complex*? Or perhaps it should be called a *Pollyanna gene complex*. Because ever since Eleanor Porter's novel produced a character named *Pollyanna*—that name stuck in our American consciousness as well as the dictionary—it has become a *real* and very common description of a certain kind of personality: Positive innocence, easily conned, *Dupable*?

Interestingly, recent research[4] has provided a fascinating and a somewhat supportive point of view for a *Pollyanna* gene complex.

Apparently a majority of human brains *process pleasant information* in a faster and more precise manner when compared to what could be considered unpleasant and/or disagreeable information. This kind of repeated perception of pleasantness would certainly have a modifying influence on human brain components. *Pollyanna* when viewed in this context could certainly be used to refer to a particular brain activity. Perhaps a group of *Pollyanna* neurons could account for the faster recognition of pleasant stimuli. Or, how about a *Pollyanna* complex-brain-*nodule* that allows pleasant *stuff* to <u>*perceptually*</u> occur more *r*egularly to us in our humdrum world. Therefore some of us have a proneness to automatically expose ourselves to what is usually defined as *positive stimuli* more frequently or even <u>*interpret neutral events as positive*</u>. CISTO! These kinds of brain *nodules* would also tend to assist us in recalling what is perceived as *to our liking* more readily. They almost certainly are the reason some people account for *statistical correlations* as *miraculous causations*. Furthermore, many a *Pollyanna* will explain such correlations as paranormal in character.[5] "The phone rang just when I was thinking of him. OMG it's a supernatural miracle!" If you see what some others see as irrational, you may roll your eyes here.

Then too, basic evolutionary survival suggests that cognitive processes, which give rise to language and behavior may, especially as children, favor pleasant/positive information over unpleasant/negative information. YEAH! Obviously, saying, "There, there—it's going to be alright." has better survival value to a child in crisis than "OMG, we're all going to die!" In addition, when parents and cultural icons reinforce pleasant/positive data our *belief* in positivity has additional clout and tends to stick. The result—the *Pollyanna nodule*—becomes a pervasive *fact* of our human existence. This also bolsters the dualistic idea that *reality* consists of two basic opposing elements, i.e. spirit and matter—mind and body—good and evil and of course, positive and negative.

Currently nearly everyone understands the term *dualism* (from Latin *duo*) to mean a *co-eternal binary opposition*. This meaning is preserved in most metaphysical and philosophical discourse but of course, as with most words it has been diluted in common usages.

Children become adults who automatically accept a dualistic view of life. They want to hear no evil, see no evil, speak no evil, and believe no evil! (However, Evil spelled backwards is Live! Inhaling requires exhaling.) Eh? Well, think about it, can we separate tears and joy?

OLD "NEW" THOUGHT

When Ralph W. Emerson published his *Nature,* (1838) extolling *new thought* and *new worship,* it was hailed as the first American self-help book. He invited

Americans to "*conform their life to the pure idea in . . . mind*." A whole generation of young thinking Americans took up his challenge of *New Thought*. Among the new thinkers was Phineas Quimby who 'taught' Mary Baker Eddy of Christian Science fame. Among their progeny were Ernest Holmes, Emmett Fox, and a dozen others. Science of Mind and related organizations in which various declensions of *Positive* were and are core words began to take shape. *Positive thinking* was and is touted as a way of life and today has become, for some popular preachers, the *science of getting rich*. Again, the emphasis is on the dualist concept of positive thinking vs. negative thinking.

Unfortunately, sometimes, the emphasis on *positive* can result in a denial of our own realistic pain. Further, when we deny our own pain, it's only a micro-step to the denial of other's pain. "Walk it off!" "Get over it!" or "Suck it up!" and/or finally, "Keep a stiff upper lip and just look at the positive side!" become the all too common verbal mantras. Some may think that to do otherwise is to enshrine *hurt* and wallow in self-pity but that's another extreme. The acknowledgement of pain as well as, when appropriate, taking time to mourn can be a crucial step toward healing.

However, in today's *cults of positivity* the pleasant view that rose-colored glasses bestow may also provide the pathway into deep ca-ca. For example, some rather famous *positive thinkers*, while quasi-professing to be "prognosticators of the future", led many of their supporters to an unrealistic view—optimistic/positive—of their prospects in the recent housing market debacle. Their oft repeated upbeat tune "God wants you to be rich," led many a *nodulized* optimist down the BS-strewn path to bankruptcy. (BS here means *bad science.*) In similar contexts politico/religious mythologies arose in which optimistic *hope* and *real expectations* are muddled. Modern examples are abundant: The best seller (19 million copies world wide) *The Secret*, promised that once you discover the *Secret* you can carve out the kind of life you want, get out of debt, buy the house you want, find a fulfilling job or even fall in love. You can have it all! In the meantime coral reefs are still disappearing, the rapid loss of arctic sea ice continues and the effects of population growth and *ethnic cleansing* bring horror and devastation to whole countries. No matter how positive our outlook or how many *New Thought* positive prayers are offered, the long-term ramifications of oil spills and pharmaceuticals are still being exposed. We saw the pitfalls of excessive optimism via the *quick war* of George W. The election of a black *hope filled* president in white America offers still another kind of optimistic possibility.[6]

When predicting future events, especially those that have devastating potential, there is no room for illogical *wishful* thinking. But then, why are we so often hoodwinked or *duped* into trusting that if we *believe we can* or hold a *positive thought,* all would be well? When such optimistic thought processes are pronounced, "proven by scientific research", and then subsumed under the rubric Science of the Mind or just plain ol' Power of Positive Thinking, congregations tend to fall in line. But in the final, underlying analysis, generalizations of this kind represent the language of *uber-belief* not a scientific theory; as such they can only be believed or disbelieved—not supported or refuted by empirical research.

THE NEED FOR *POLLYANNA NODULES*

The words *duped* and *hoodwinked* can mean the amalgamation of optimism and naivety, which is also a pretty good description of *Pollyanna*. As said earlier, if there is a *Pollyanna* gene complex it must in cahoots with a *dupe gene complex*. Years of listening to *duping* would undoubtedly result in a huge brain *nodule* to boot. As we have already determined, when self-consciousness arrived and we primitive hominids became aware of our own mortality (and that we were mere transient critters), we were undoubtedly frightened in the extreme. After all, we had to now be aware of and face our own death! In order to survive the anxiety engendered by this *arriving insight*, mechanisms were gradually selected into our species genotype that helped to reduce this painful unease. Our amazing neo-cortex (especially prefrontal lobes) allowed, perhaps compelled, our early ancestors to dream up various futuristic pain free realities. In this alternate future world they imagined themselves existing beyond physical death. That was and *is* a '*comforting thought*' and undoubtedly has 'survival' value!

It also seems obvious that we humans love to (probably need to) participate in *duping* ourselves. We are enamored of all kinds of magical and illusionary thinking via our evolutionary history and our very real need to deal with our fears. Following the fruition of neurological s*el*f-consciousness, Homo sapiens continued to evolve new needs and *invent* still newer ways to deal with personal mortality. The existential angst resulting from the inevitability of *death* needed and received a solution. That frequently repeated solution, (in home, school and office) no doubt, produced a magnificent *nodule.*

The fact that every culture from the dawn of history evolved words and *uber-beliefs* regarding so-called alternate realities and engaged in life-after-death rituals that advocate a belief in a *heavenly paradise* are fundamental ingredients in our genetically derived self-*duping* process. (Archeological burial evidence demonstrates that even the early Neanderthals had such rituals.)

Today, I may pay money to have magicians trick me. In all likelihood *Pollyanna genes and tributaries* are responsible. I along with most of my friends love Santa Claus in all his forms and new metaphysical constructs are constantly being invented to appeal to what must be our human *Pollyanna gene complex*—but then consequently *dupe nodules* grow and take over. Free choice goes out the window and guess what? We have more and more hardcore *uber-believers* willing to fight and even die because in their Pollyanna *belief system* a metaphysical paradise awaits! GMAB!

During a visit to the Temple of the Dead in Greece some years ago, the archeologists working there invited me to view some of the *duping* mechanisms the wizards of Homer's day used to demonstrate magical object lessons to their Pollyannaish congregations. Whole walls would move out of sight and apparitions could be made to swing into view from vaporous clouds hanging about in the side wings. Slight of hand, disappearing walls and visual aids of all sorts helped the Hellenic priests to provide a convincing show of what the Olympian gods desired for human beings. Similar constructs have been elaborated on ever since.

Modern dictionaries suggest that if one is *duped* or possesses *a Pollyanna* temperament, one will tend to function as the *tool* of another person, or will be easily conned by religious or political systems. Considering that half of the world's population is below average intelligence and therefore likely to have less formal education than the other half, it seems reasonable that this *lack of education group* would be more highly correlated with the *Pollyanna* process. Yes, right again, it has been found that education is inversely correlated with authoritarianism. That is, Chris Brand's research[8] shows that the more education one has, the less need one has for authoritarian direction.

IS IT RIGHT TO DUPE A CHILD?

Children especially love fantasy, magic and illusion. The mystery of how Santa Claus can slide down a chimney and deliver all those glitzy presents on Christmas morning or the enchantment of an Easter bunny hiding eggs in grass, keeps kids enthralled for hours. Interestingly, many parents seem tentative when confronted with the need to reveal the speciousness of Santa or the Bunny and especially the Tooth Fairy. It seems that for many, to lose this *sense of magic* is to lose innocence itself.

Since this *love of magic* begins so very early in life it must be totally genetic rather than learned. Oh? . . . Well, think about this: Infancy typically is filled with cries for help (food, comfort, protection etc.) and when help comes in the way of food, comfort or the miraculous presence of the *cooing mom creature* it must seem like magic. Even as adults, we are all so in awe of multi-tasking mothers.[7] They work at 9-5 jobs, prepare meals, do the laundry and care for a couple of kids; they are doctors, mediators, and cab drivers. They do it on a daily basis, year in and year out as if by magic, all at the same time. Talk about enchantment and *Pollyanna*! Of course, what goes on in the mind of an infant is anybody's guess but I think it must seem like the '*supernatural*' at work. We can only surmise that the parental all-powerful provider seems omnipotent and omniscient to the maturing child. Magical for sure! The lucky child then, lives, grows and evolves within a kind of *magic kingdom* as naturally as a fish in water. (Do fish thirst?) Needs and desires are understood and fulfilled in a twinkling. Is it any wonder that children resonate to fairy tales, ghost stories and supernatural lore of all kinds?

But from a genetic point of view there have been innumerable s*urvival bottlenecks* in our evolutionary history that would also allow for the selection of genes to promote *duping* and *Pollyanna thinking*—even for children. In times of extreme, almost impossible, circumstances individuals who are *duped* into struggling on, expecting magic and miracles might survive, while those who were not *Pollyannaish* would just give up and perish. The survivors, of course, would pass on their genes. Such circumstances appear to be the beginning of what we often call *hope*. Thus, as we human beings are born, develop and grow to adulthood we see that both environmentally and genetically we are programmed to become *hope* filled beings. YKWIM? This is across-cultural worldwide phenomenon.

In the 50's a researcher name Curt Richter[8] at Johns Hopkins University conducted a number of what some have called sadistic studies on wild rats. In one study, Richter held the rats in his gloved hands in such a way that even when they tried to wiggle, his tight grip prevented movement. After some time, the rat would cease to struggle and lay still. Richter then put them in a basin filled with lukewarm water (deep enough so the rats could not touch the bottom and too high to climb out). Although the rats were capable of swimming, they would not even try to swim, they just gave up and drowned. Then in another study, Richter took wild rats and put them directly into the basin. The rats were placed into the water and then observed. Within 45 minutes all of the rats had just given up and drowned. Next, the test was duplicated and just as the rats were about to give up, they were rescued. Then the same rats were again put into the basin and the same rats were able to swim for upwards of 60 hours before giving up! It was said they had *hope* of some magical intervention.

This may or may not suggest to you that when an infant is *rescued* thousands of times in the process of becoming a child and then thousands of more times while approaching adolescence, this same *hope filled* person is *duped* into becoming a *Pollyanna believer.*

Magic? I think clouds are magical as well as moonlight shimmering on the ocean. Even now as I write this, the mountains surrounding Yosemite are awesomely magical. When a wild animal pops out here or there that in itself is a magical moment. Right now it's a magically bright sunny morning. Green IS green and blue IS blue. More magic! I'm getting ready to head up to the Napa valley for a magical week. I have been told that Bastille Day is also magical in certain quarters of the Valley. Anticipation and fantasy is half the fun and magical in and of itself. I know there will be great wine and superb food!

Perhaps now you agree with me that the word *magic* can be over-used. But think about it: What other words can we draw on? Where do people go for reassurance? Ah yes, there is the word *hope* to fill in the blank. It may be that people who ponder some of the questions we are discussing in this book are more anxiety ridden than those who never take time to consider such matters. Yet, as a friend of mine once said, "it sometimes seems that nothing is offered to a questioning person beyond a bleak future of crypts, coffins and an unfathomable eternity. They may keep searching because it seems that the universe is so immense and so complex that our hundred billion minuscule brain neurons can't possibly be the full answer for a *hopeful* human being. This life that we all possess, surely it can't just be *luck*. There must be some kind of *magic* somewhere that is huge and incomprehensively powerful." Undoubtedly such thoughts do bring us, and especially children, comfort. And here is where historical '*hope filled*' ideology justifies itself. Amid the arguments made by new anti-religious writers such as Richard Dawkins, Sam Harris, and Christopher Hitchens, the inspirational writing of Karen Armstrong makes the case that the religious life can be valuable and healthy. (In terms of anxiety reduction and mental health.) While perhaps not necessary for the objective reality that is essential for life itself, religious ideology, helps many "to live creatively, peacefully, and even joyously with realities for which there [are] no easy explanations and problems that we could not solve [i.e.] pain, grief, despair, and outrage at the injustice and cruelty of life." Yes, comfort and anxiety reduction appear to be valid reasons for some people to *believe*. It seems justified, to have a mother say to a child: "It's going to be alright!" That sentiment (true or not) is an essential part of parenting. However at other times, the <u>reality</u> is, "It's not going to be alright" and perhaps we <u>as adults</u> need to be able to face that reality as well. Furthermore, once we embrace the *harbor of consciousness*—the brain/body itself as the sine quo non—and begin to experience the ongoing flood of miracles that neuroscience and genetics now offer, we can be astonished as well as become calmed by the "limitless promise of ever-unfolding-newness" being revealed. Magic indeed!

LUCK, MAGIC & THEOLOGY

As a ponderer (curious and questioning) person <u>you</u> can certainly use the word *magic* or even *super-natural* to describe the trillions of miraculous minuscule cell connections in <u>your</u> own two and a half to three pounds of neurological workings that perceive, learn, intuit, forget, and re-learn. No problem! But it is interesting to note that words like magic, chance, luck and supernatural are found in every language and are almost always used to describe so-called incomprehensible *events* that seem to occur regularly by accident or chance all around us. In 2009 an amateur treasure hunter stumbled across the largest hoard of gold coins ever found buried in England. Dave Crisp, who was said to be one "lucky dude", discovered the ancient stash, valued at more than $5 million. However, kids in Cambodia regularly lose their legs and even their lives as the result of land mines buried during the Viet Nam debacle. This may be called a misfortune or even an accident or were they just "unlucky"?

When playing fetch with my dog Mr. Dugan I throw a ball with deliberation and intent. The ball comes back to earth not by chance or luck but by what we understand as a loss of inertia and gravity. However, the ball may return to earth and strike a beveled wall and squirt into a fishpond. OMG! That bit of mathematically predicable physics can be called bad luck or accidental because it happened without my intending it. We may say that the ball wound up in the pond due to an unfortunate conjuncture of causes. The philosopher Aristotle attempted to deal with such chance accidents by defining luck as the "incidental production of some significant result by a cause that took its place in the causal chain incidentally, without the result in question being contemplated." (*Physics BK II Ch. V*) And voila!

Because of Aristotle's next conclusion, Theologians and Philosophers ever since have been injecting a *supreme being* as a causative factor. But, before we get to that, let's propose that two physicists or two bicyclists collide at a blind intersection at a particular moment in time, and the 'cause' is called accident or bad luck. But since this kind of hypothetical accident was unforeseen and unintended, Aristotle would suggest that the conjuncture of the two bicyclists was incidental to the intended purpose of each rider and was therefore beyond their control and thus just bad luck.

However, this should not suggest that in this day and age of GPS and portable radar-like warning systems there could be no comprehensible statement or predicable data available to control such *accidental effects*. Today, Aristotle's pronouncements notwithstanding, this *accident* could have been predicted with a great degree of certainty and would therefore fall out of the purview of the gods. JMO (*just my opinion*) And that is the point! Every decade has its unfathomable mysteries and every decade has new scientific solutions.

Some philosophers and theologians have suggested that uncertainty and lack of constancy have traditionally been the hallmark of events that happen by chance or accident. Abracadabra, utilizing Aristotle's logic, they say it follows that an event that occurs regularly and constantly according to the laws of nature cannot be a function of *chance* or *accident*. They conclude, using philosopher A's logic, that since the universe (at least certain aspects of it) appears to be governed by law and order it cannot be the result of chance or accident but is rather governed by purposeful design. They repeat the words of Aristotle, "Nature achieves invariable results . . . That Nature is a cause, then, and a goal-directed cause, is above dispute. (*Physics Bk II, Ch VIII*) Ever since, Philosophers have been introducing God here but of course, Aristotle didn't know about GPS, radar, exploding super novas, black holes and the general chaos and quantum-ized (random) order of our ever-expanding universe. Once again the incomprehensible becomes comprehensible! Invariable random results are the core of Quantum physics, but no reputable physicist would call on God as an explanation.

ALTERNATIVES

So, when and if we need reassurance about our eternal future we can ponder the many metaphors philosophers have passed down through the ages (e.g. "the kingdom of God is within") and find great reassurance that truly and according to the best science: "Nothing is lost in the universe." Every day and every question in life needs to be explored to the fullest with glee of a Voltaire[10] or a Twain or Mencken. There is not one of us who does not possess the ability to participate in this kind of daily exploration. True, it will always be a matter of integrity and commitment. But . . . if . . . as you may have guessed, Neuro-nastics (REHAB) is considered, it will facilitate the modification of *nodules*. It *will take effort and time to see results, but* in the meantime we can enjoy to the fullest the many metaphors that bring illumination to the great questions. That too, is *magic*!

Homo sapiens intellectual development has been, with few exceptions, an eons-long struggle to observe happenings and report them to others in the most comprehensible way possible. And yes, metaphors can be useful! But when there is (for the moment) an incomprehensible event, these same *Homo sapiens* are not above utilizing their highly developed *dupe nodule* to project their own internal fantasies and emotions outward. For example, interpreting thunder as a temper tantrum of the gods or a lightning strike as a well-deserved chastisement for sin. These kinds of projections are not just ancient history. Even in the 21st century, some leading *dupers* have named cancer and AIDS as divine punishment. Notwithstanding such mind projections, in more recent months there appears to be an ever-dawning understanding that the world's happenings are unfolding according to probability ratios, natural cause and effect, and discoverable algorithms. None of which have the slightest connection to our capricious human emotion or projections of magical otherworld dimensions. Many who previously acknowledged themselves as atheists, and agnostics are now calling themselves *Naturalists* as contrasted with *Super-naturalists*.

I agree with Rousseau who said, "<u>Nature</u> never deceives us; it is we who deceive ourselves". CISTO *(can I say that out loud?)*

CHAPTER 3

FOR THE LOVE OF SILLINESS

Correlation VS causation

Like some exotic modern disease, a contagion of *silly ideas* appears to be the price we, freedom loving humans, pay for our diverse, comprehensive, socialized existence. In part, because *we* have such a voracious appetite for answers, we will entertain *uber-beliefs*, even if the answers reek of absurdity. Ancient neural hard wiring probably insists that we find some answer—rather than none—in order to attempt a reduction of mental gridlock. But as a consequence ill-logic seems to thrive in every hominid gathering place because some of our complex multi-ethnic societies have come to depend on mythical tales or the interpretations of a few charismatic *dupers*. These *dupers* can, with a good deal of sincerity and eloquence, make silliness and *uber-beliefs* appear as *ultimate truth*—especially to those of us who are at heart, *Pollyannaish* and think so optimistically about everyone and everything.

Ah yes, we human beings, despite centuries of progress, education and democratization, appear to still be hard-wired to sort ourselves into biased pecking orders within clans, clubs and political parties, expecting—even requiring—ourselves and others to adhere to explanations (no matter how irrational) that calm us. Recently, Neuroscientist Robert Sapolsky and his associates at Stanford University in California shocked the academic world by linking human perception of *skin color,* to the foundation of racism. For example, the *Amygdala (Part of the limbic system—small almond-shaped structure in the anterior part—plays a vital role in emotional behaviors such as fear, anxiety and aggression.)* lights up when a scary looking or a different color face is presented, even subliminally, to Caucasian college students. Sapolsky said, *"Damn, that's an upsetting finding."* Once again our neural structures, rather than *choice* come into view as an instigator of an alternative explanation for racism. OMG!

Like other herding animals, we huddle together in urbanized corrals usually depending on the politically select few to inject their prescriptions and provide a tract to follow. Yes, absurd as it may seem, we love our politicians—we adore our stand-up comedians and most of the time we strain feverishly to tell them apart.

Our *need to follow our neural hardwiring*, as well as be *duped into a calmed down state* appears to be implicit within our human genomic structure. We build our environmental and institutional thought composites using *Pollyannaish* ideas as the *bricks* and myth as *mortar*. Like stiffening rigor mortise we accept a *fixed sheep-like mode* as our idealized diagnosis for following the leader. More often than not, a leader's ideas are merely an attempt to mollify the group's emotional dissonance and thus provide the flock a collection of prescriptive and soothing answers—no matter how silly. Usually they come bundled together in response to questions about seemingly inexplicable feelings, events or dis-eases. We almost swoon when told: "it is a mystery!" We love magic and magicians even while knowing we are being *duped* by their illusions. But, in this same context, that which is an emergency—emerges. Sudden political shifts or *new* hitherto unnoticed environmental or technological happenings[1] can and often do create *new* kinds of societal stress and thus a critical need for modified answers that soothe and comfort. Those on whom we depend for relief (gurus, politicians, and comedians) interpret the possible meanings of rising stressors and fit their answers into recognizable *belief systems*. Thus silly ideas become ritualized and publicized—voila! A solemn ceremony is passed on to the entire assemblage. If it is practiced on a full moon night or on the stage at Kodak Center it, more often than not, becomes entertainingly addictive. Our brain *nodules* grow accordingly and suddenly it is a monumental task to find our way out of the ritualized but mythical labyrinth.

Interestingly, the proffered interpretations are often explained in such a way that they cannot be checked for authenticity. One of my mother's favorite retorts, when confronted with some of my impossible childish inquiries was: "God only knows!" There are dozens of others: e.g. "It is Allah's will"—"There are more things in heaven and earth Horatio . . . ."—"it must be taken on faith . . ."—"Fifty thousand Frenchmen can't be wrong!"—"A man and a parrot went into a bar."

This is not to say we cannot, if we are inclined to do so, assess the logic, and/or silliness of ideas as well as the reliability of a leader. However most of the time human gullibility and its subordinate *faith in* BS (*bad science*) play out their *nodulized* roles—the result: *absurdity and silliness rule the day*. Like the story of the ninety-five year old man who proudly proclaimed to his doctor that his eighteen-year-old wife was pregnant. Doctor listened and then said, "A man went hunting for bear. Absentmindedly, instead of his rifle, he grabbed an umbrella and headed for the woods. When a bear charged he raised his umbrella shot the bear and killed it."

The older man said, "That's silly! Someone else in the woods must have shot the bear."

The doctor replied, "my point—exactly!"

There is little doubt that simple analogies or double-checks of logic could and *sometimes do* protect us from a whole lot of nonsense. Unfortunately, these checks are far from perfect and seldom used, especially when confronting an *uber-belief.* As a result it seems we cannot escape certain types of *absurdity.* Recent worldwide events seem to corroborate this.[2] (e.g. racism, mindless terrorism, WMD searches, ethnic cleansing, etc.)

CORRELATION AND CAUSATION

Correlation usually suggests a relationship between two different items or experiences (called variables), but they cannot prove that one variable causes a change in another variable. As much as we would sometimes like to believe otherwise, correlation does not equal causation. A poplar example: correlation might suggest a relationship between a straight 'A' student and self-esteem, but it cannot show if academic success increases or decreases self-esteem or vise versa. Other variables might play a role, including, wealth, social standing, I.Q. and/or a perfect set of teeth revealing a dazzling smile.

Another, perhaps more obvious example, comes from the observation that sleeping with your shoes on is highly correlated with waking the next morning with a headache. Therefore—cum hoc ergo propter hoc—sleeping with shoes on causes headaches. But, should we consider the two bottles of cheap Champagne you drank the night before? YKWIM?

Here's a beaut: As snow cones and ice sales increase, the rate of child deaths by drowning increases sharply. Therefore, snow cones and ice cause kids to drown. Duh? Snow cones and ice are sold during the hot months at a much greater rate. It is during the summer months that children are more likely to engage in activities involving water, such as swimming or playing around water. The increased drowning deaths are simply caused by more exposure to water based activities, not ice or snow cones.

The reason we do repeated empirical studies in the first place is so that our *Pollyannaish* observations and biases don't hoodwink us. Take the claim that some people have telepathic abilities. There is a million dollar reward waiting for such evidence. * Have you asked, "Why hasn't it been claimed?"

Personally, (I am very cautious about this) I have a distant *'idea'* that if we can release current fairy-tale *'beliefs'* and allow brain evolution to move onward, we will, in some distant century, create something like communication via telepathy but for now, notwithstanding fringe science's non-peer reviewed assertions, ESP is still woo woo non-science. (Unless individuals are *randomly assigned* to participate as receivers of so-called ESP messages, the results of will forever be suspect. See story of Hans the horse in Chapter 5)

Few things mislead us more than failure to grasp simple statistical principles such as regression to the mean—basically, returning to the average. We can be fooled for example, about why we recover from an illness after receiving a weird home remedy or even a treatment provided by traditional modern medicine. Most common ailments including head colds and generalized pain have what we could call natural cycles. We tend to seek treatment when the symptoms are peaking. This is also where the regression to the mean would suggest that we are most likely to return to our average or usual health condition. When we recover we invoke *Pollyanna* and swear by the treatment. (If our regular treatment is antibiotics for a cold or stuffy nose we will more than likely facilitate a practice that will eventually make possible the development of drug resistant bios that will kill us.) Good luck with that one!

What we need now in this time of proliferating quack science and instantaneous communication is a BS (*bad science*) detector. Just a few comments will suffice: "Show the evidence!"—if the response is personal observation, intuition, or anecdotal, IMO you can be 99.99% sure it is wrong. Or better still: please replicate the happening—with laboratory or other environmental changes. Carl Sagan used to say, "Extraordinary claims require extraordinary proof." GS (*good science*) is not just a collection of observations or even what we call facts. Good Science is a method of repeatedly and incessantly interrogating our whole world.

A PARADOX

In America, philosophers, standup comedians and most social scientists struggle with the frightening and paradoxical fusing of our impressive intelligence and our stunning stupidity and willingness to be duped. Even more frightening is the realization that the understanding of this paradox is becoming critical to our survival. Americans are smart enough to invent technology that can, with the push of a button, destroy our planet and at the same time someone on this planet may be just stupid enough to push that button. Americans are smart enough to assemble the political-cultural and technological solutions for today's challenges and so very dim-witted in the utilization of them. On one hand we develop the capability for instantaneous worldwide communication, on the other, our politicians actually boast of their <u>unwillingness</u> to have an open discussion.

INTELLIGENT ADAPTABILITY

Almost certainly we have survived thus far because of another genetic endowment known as *adaptableness*. But this is an ancient aptitude rather than a modern one. *Intelligent adaptability*[3] allowed hunters and gatherers of prehistoric times to survive. That same type of intelligence allows the survival of traditional indigenous peoples today from the bleak Mongolian plains to the forbidding Australian Outback. Arctic people and isolated Pacific islanders succeed in difficult environments because of their *intelligent adaptability*. Even at this juncture in modern time, most of these peoples *believe* that their gods and various kinds of supernatural enchantments enable them to survive in their dangerous worlds. Their *belief* in demons, hell-fire, witchcraft, totems, omens and evil spirits may seem silly to outsiders from the modern world and yet perfectly understandable as an ancient cultural phenomenon. Such beliefs in hostile primitive settings undoubtedly contribute to comfort (stress reduction) and thus survival.

Yet, in our modern world this *ancient intelligence* is still being combined with our modern *high tech intelligence* simply because technology has out stripped human cortical evolution—that is, the ever changing human brain has not entirely left behind the need for replacing factual data with *uber-beliefs*. We should remember however, that such *beliefs* are characteristically a response to misunderstanding, lack of data or social chaos.

Both indigenous traditional <u>and</u> sophisticated modern peoples are tormented by fears related to superstitions, evil spirits, witchcraft and mythological threats of an eternal hell or loss of a future eternal paradise simply because our *Pollyanna* brain has not yet developed beyond. The human historical record documents a paradoxical mix of *insightful survival* and inconceivable *irrationality*.[4] This combination has carried over to our present day but with the *tech zeitgeist* of our modern existence, insightful survival epiphanies have not kept pace. Our prefrontal anticipatory faculties have not advanced fast enough to allow for our continued existence on this overcrowded and polluted planet. The challenge of today's rational person is to explain, even to a child, how we can concurrently be so bright and so dim. Of course, if this were not the case, Cris, Robin and Whoopi would not be in the comedy business very long. YKWIM?

Then too, our modern intractable silliness would not be so noticeable or much of a problem if it were not so costly. So much time and energy is wasted within our contemporary metropolitans; our politicians trap themselves in a mire of gridlock, fighting each other over turf and power while publically offering protection from imagined dangers to gain the media's and thus the public's attention. This represents the *ancient intelligence* in action. *Defense* and *survival* meant everything (and still does) to those of the *ancient intelligent* mind set.

Since 9/11 irrational political, and theological *uber-beliefs* have cost tens of thousands of lives. Again, the absurdity would not be so much of a shock if we lived in that ancient time when details about the faint features of comparatively unknown worlds were not accessible. But today, information evolution has outstripped our physical evolution. We can now see and understand the subtlest aspects of our universe—both microscopically and telescopically—they are both at our fingertips. Furthermore, what the telescope did for astronomy and what the microscope did for biology, neuro-scans have done for the totality of modern brain science. Yet large segments of our civilized and educated populations remain fixated in the extraordinary absurdity of the *ancient intelligence* with all of its *uber-beliefs*, incumbent fears and superstitions.

Both Eastern <u>and</u> Western peoples find themselves at odds with modern scientific thinking. Why? It would appear that even more than primitive peoples, large segments of *Americans* love being deluded and manipulated. From Las Vegas conjuring to a myriad of dramatic TV programs (misleadingly called *reality*) we pay with funds as well as our time to be tricked. Why does such silliness endure? Can an ever-developing understanding of human intelligence be reconciled with these currently ongoing and costly irrationalities? So true, TTTT.

A friend of mine suggested that this book be called, "Why are we so mean to each other? Why?" I thought about that and realized that if we address the more generic query: '*How can we reconcile the contradictory mix of our incredible intelligence and our profound irrationality?*'—*an* answer will no doubt emerge.

ADAPTIVE INTELLIGENCE

Please note, today's Homo sapiens's success has no historical precedent. In the same breath we must admit that today's worldwide horror (typically triggered by silly-minded *uber-believing* East and West Homo-sapiens) also has no historical precedent. Even the Crusades and the Inquisition pale by comparison to today's worldwide terrorism and the resultant wars.

The immediate basis of most of our contemporary successes may be tagged: "our easily modifiable brainpower"; that is, our remarkable talent to *think ahead and act creatively to invent and evolve* contented life styles. In fact, we Americans have often been derided as lacking in <u>culture</u> and accused of a national ADD (*attention deficit disorder*). From another point of view this can be seen as part and parcel of America's success. We change our political paradigms as quickly as citizens of other nations change their socks.

We hastily, almost heedlessly, utilize slick technology and financial means to decipher and solve the problems thrown at us by modern life as well as Mother Nature. Today in America, comfort, not survival, appears to be the ultimate aspiration. However, an ancient aptitude, (adaptability) coupled with self-consciousness and our stunning pre-frontal cortex, with which we are able to think about the future, creates an advantage over other life forms. This complex has been used to our distinct benefit since our written history began. At the same time, we hate to admit, it seems we fear the resultant freedom our *adaptive intelligence* allows. Thus instead of facilitating human development we devolve to *ancient mythical ideology* (*uber-beliefs*) and the *limits* it places upon our thinking as well as brain change. Our ever-evolving creativity is mitigated by our silly irrationality, which diagnoses our fear of freedom and a lack of progressive Homo-sapiens brain training. Instead of new epiphanies to convey peace we have new materials to build missiles. Pretty primitive stuff!

OTHER CULTURES

As the year 2000 approached I went to the Far East to experience what popular media called "the birth of a new century." The budding of China as the world power was obvious. The new ruling elite, with their Mercedes Benz and Armani suits no longer feared to be conspicuous. Years ago China had, more or less, withdrawn from international contact. Then, she managed to by-pass the costs of more than twenty years of high tech research and development. China emerged from her self-imposed isolation and suddenly all the technology other nations had so tediously and expensively developed was at her immediate disposal. Old Beijing and new Shanghai are both virtually free of telephone wires. Almost all of the men and women riding twelve million bicycles or electric scooters carry cell phones. Cell phones are as common as the Mao jacket used to be. Somehow China has been able to reconcile the absurdities of Communism with the benefits of Capitalism.

My *twenty-first century* journey took me north to Mongolia and there I saw vigorous people living as they had lived for thousands of years in cozy Yurts with the famous Steppes-ponies as their primary means of transport. On one hand, these virile tribes-people are still legendary for their *adaptive intelligence* and their ability to survive and live as one with their severe environs. On the other, they appeared to me to be plagued by superstition and silly mythology. Again, pretty primitive stuff!

WHY ABSURDITY PERSISTS

There are many explanations for the survival of our absurd thinking and the lack of continuing brain evolution. Here I offer a few:

1. First, and least satisfying, is that *belief*—in some super something, no matter what—brings comfort. (Anxiety reduction) The educated, affluent, young men who flew into the towers on 9/11 were undoubtedly comforted by their *belief* that they would be received into paradise with all its attendant pleasures. Over and over I hear people say they *believe* as their parents *believed*

because it brings them comfort to do so. Best sellers are still being written that have as their core message: "*delusion,* if it brings comfort, is desirable."[5]

2. Evolutionary biologists have offered a second and completely different reason for human irrationality. It is encapsulated in the mantra of *'change'* and our willingness to 'leap before we look'. The beautifully feathered birds of Hawaii have all but disappeared under the onslaught of the Mongoose introduced from India to kill off the sugar-cane rats, which probably came from England courtesy of Captain Cook. The rats hunted by night and the Mongoose hunted by day thus they avoided meeting each other. So the Mongoose turned to easy prey. Within a few short years all of the gorgeous ground birds, whose feathers were used in the headdresses of the Hawaiian Royalty, were gone. The Mongoose thrives!

Latching on to quick answers to produce change can lead to irrational action as well as thinking. A more recent example is the yearly seven billion dollar subsidy to corn growers to produce a second rate fuel called Ethanol. Even though there has been no reduction in foreign oil imports and Ethanol has proven to be destructive to fuel lines and actually pollutes more than gasoline we still insist on producing it. Yes, it appears that we still continue to "jump before we actually look", especially when prompted by *dupers.* It was America's own home spun Walden Pond philosopher Thoreau who said, "It is a characteristic of wisdom not to do desperate things." In his day he was called ACORN (*A completely obsessive real nut*). He also said: "What is the use of a house if you haven't got a tolerable planet to put it on?"

3. Recently, neuro-researchers have come up with still another explanation for our silly ideation. Neuroscientists tell us that the human brain, even while using only a fraction of its computing power, while making plain ol' ordinary decisions, deciphering the simple, mundane changes incumbent in everyday life, still sops up incredible amounts of data. And because it all happens so seamlessly, most of us find it hard to realize that in making our *quality decisions*, billions of neuron connections are receptive to myriads of subtle cues and use constantly updated information. This complex and near picture perfect process is then flawed and made more difficult—i.e. neutralized or BO—by our *nodulized* biases. (e.g. *uber-beliefs* like Ethanol will save us!) Instead of accepting our new '*quality decisions*' we tend to retreat to familiar BO (*beiged-out*) habitual beliefs.

To avoid being exploited by bogus explanations, or fixated by previous indoctrination (i.e. *nodule* growth) we need to be able to step out of our '*status quo comfort box*' and trust our new input and our *new quality decisions*. Only as we extend our *thought-parameter-trust* are we able to read the motives, intentions and emotions of others, especially those who constantly tout faith or hope as the panacea of society's ills. Luckily, it is still possible for us to use the discerning part of *adaptive intelligence* to overcome *micro-nodules* as well as the *silliness* that is passed off as an explanation. As a matter of fact, we respond slightly more rationally and effectively when we are confronted with problems requiring *adaptive intelligence*. Most scientists believe this is because we have evolved what could be called *adaptive* brain-systems capable of reading environmental cues as well as the intentions of others. It has been suggested *that we behave intelligently when we are confronted with predicaments for which we have developed particular capabilities*—language, trade, symbols, social judgment, simple tools and survival itself—provided we have not already developed too many oppositional *nodules*. For instance, most of us have observed individuals who become irrational when faced with new, novel challenges especially if modification of a well-developed brain *nodule* (*uber-belief*) is required. It would seem then, that we could readily meet a challenge of change if the problem looked somewhat similar to a problem of the world from which we have emerged.[6] Again, the realization, that we are constantly looking backwards to solve current challenges and utilizing u*ber-beliefs* instead of seeking up-to-date scientific solutions to current crises, could mean that we have somehow not allowed continued brain/mind evolution and thus we are deprived of new cognitive abilities. It seems then, that perhaps we could all do with some REHAB-NEURO-NASTICS to overcome our fixation with the past.

Even so, for us to accurately judge the sincerity of modern TV personalities (politicians, evangelists or motivational gurus) via their million dollar mass media ads is near impossible. To judge the sincerity of a common Email IS often impossible except as the label *Spam* enlightens us!

It seems then that we become more comfortable with today's problem solving, to the extent that our modern challenges resemble the ancient ones from whence our so-called *intuitive talents* evolved. We fumble, fail and appeal to absurdity and *uber-beliefs* when we are confronted with *change* that appears unprecedented or is in conflict with our already developed brain-*nodules*.

4. Still another rationale for human absurdity is found in the *conflict* arena. It is hard for most of us to admit that the major threat to human existence is the irrational conflicts born out of silly thought processes. Since we live in groups and have done so for eons, we have learned that to cooperate with others is a necessary ingredient to survival. Cooperation is most probable when the interests of both parties overlap. But here is the rub: the interests of even two human beings seldom, if ever, completely coincide! Even the interests of identical twins are not identically identical.[7] Thus, implicit within cooperation is the dynamic of disagreement. Conflict and cooperation are the Yin and Yang of civilization's

continuum. As the old song says, *"Can't have one without the other."* The human brain recognized this long ago. Conflict over the spoils of a hunt was implicit in the joint venture of hunting down Mammoths or harvesting wild berries. My friend's question—"what makes us so mean to each other?"—has a partial answer in this context. From ancient hunting collectives to modern condominium associations, <u>conflict has been a relentless escort</u>. It is part and parcel of cooperative endeavor. How to equitably divide the resources of a joint venture is the bottom line for most modern political parties. Civilized legal systems, if you care to look at the bottom line, are seeking a fair and balanced way to divide the spoils. Cooperative ventures—societies, large or small—generate new assets and thus new grounds for inevitable conflict.

Lack of awareness of this seeming paradox can generate heated disagreement and yes, even *downright meanness and war.* In contrast, awareness can create a mutually beneficial trade off. A while back I was in a furniture store and observed what appeared to be heated haggling over the price of a mattress. As I listened to the seemingly inane strife, I became aware that the two men were bargaining in a mutually understood arena of subtle reciprocal cooperation. As a result of their understanding and appreciation of each other's point of view, a handshake and a jointly beneficial trade occurred. With such awareness we can be assured that <u>wherever we see cooperation we can</u>

be prepared to see conflict. That observation should not be interpreted as a lack of collaboration, rather conflict should be seen as ubiquitously connected to cooperation. (Can you make physical love without friction?) The outcome and change of attitude of the two quibbling men in the furniture store may be a mystery to some. For these, a complex explanation may be required, but the more complex the explanation the more vulnerable listeners become to exploitation. The Law of Parsimony (*The simplest explanation is consistently closer to legitimacy.*) could be called *the solution to silliness and misunderstanding* because it seeks the simplest most unified solution.

WHY CIVILIZATIONS FAIL

The saber rattling, idiotic political threats and murderous activities generated from *uber-belief* of ancient unfounded fables is one of the ultimate absurdities of modern times. Arrogant, self-aggrandizing, wishful thinking about this or that god—granted paradise has led more than one culture to collectively self-destruct. When I read Jared Diamond's[8] *Collapse* I abhorred the absurdities revealed. He traveled to Easter Island to explore the mystery of those huge brooding Mo'ai rock statues and why the Islanders died out almost completely. The *why* is that they failed to see their own end coming because they were in conflict with each other over *silly uber-beliefs*. They killed every tree on the Island to transport their stone gods to a particular place on the same Island. With the trees gone the soil washed to sea and their crops failed for lack of nourishment. The Island clans then turned on one another and fought to the death over scarce resources. They literally killed themselves off with their conflicts over their *beliefs*.

Diamond also examined the Greenland Viking of a thousand years ago. They sailed into the steep Greenland fjords and turned the grassy slopes into sparse pasturelands for cattle. Because of their beliefs they quickly found themselves in conflict with their environment. They hunted Caribou but placed a taboo on the calorie rich and plentiful seals. They were completely united in their construction of a large Cathedral but not so much in their construction of fishing nets. Finally, starving to death, they ate their dogs but still refused to eat crustaceans. Right up to the last these faithful believers never lost sight of what they stood for! Talk about *nodules and uber-beliefs!* They died proving their point—or did they?

Rebecca Costa, in her ominous sounding book, *The Watchman's Rattle* offers a fascinating account of similar events among the Mayans. Apparently they had the technology needed to survive but instead relied on their "faith" and in magical *beliefs*.

Cambridge anthropologist JD Unwin way back in 1934 put forth still another theory of 'civilization downfall'. He studied 80 cultures over 5000 years and decided that without exception, "*cultures that practiced strict monogamy in marital bonds exhibited creative social energy, and reached the zenith of production. Cultures that had no restraint on sexuality, without exception, deteriorated into mediocrity and chaos.*"

The nightmarish poignancy of the downfall of all of the above mentioned societies could be an obvious parable for us today. Initially, we may smile at their silliness that led to their conflict and demise. Then the realization arrives: We too are in a self-delusional conflict of our own. Because of our need to hold fast to a particular ancient *belief* or to be politically correct—"not being able to LOL at our prevalent *uber-belief* oriented mythology"—we thus facilitate the almost forced acceptance of the most popular fables, and the attending *nodule* growth. The end result: We are threatened with annihilation. Today, is there a thinking person who doubts that—because of *beliefs* and the resultant conflicts—we are heading to an Easter Island fate? They never saw it coming. Do we? IGTP (*I get the point*)

Apparently many *moderns* are conflicted about the need for immediate adoption of an enlightened life-style that could make a huge difference on this planet—albeit decades from now. We talk of *carbon footprints and alternative fuels* but it may be that we, and more especially our leaders, are literally incapable of thinking in timeframes that exceed our own life spans. That may seem absurd but there is neurological evidence that supports the inability of some to anticipate or plan ahead.[9] The growth of *nodules* would facilitate this lack. Rather than continuing *heavenly comfort* as the ideal, we need to evolve and be open to new kinds of *earthly* epiphanies.

THE COSMIC JOKE

This brings us to still another explanation for human absurdity. As hinted previously, recent neurological evidence reveals that the human brain itself may be the culprit. Neuro-research[10] suggests that human absurdities in the form of divine directives and visions originate in specific parts of the brain. In other words, so-called divine epiphanies resulting in *visions or inspired writings* can be shown in laboratories to be neurologically derived. It is now possible, with the help of various high-resolution brain scans to chart the neurological pathways that are switched on and off during the performance of pious rituals such as chanting, meditation or prayer. Modern neuro-scientists want to find the sources of some of our silliness. The resultant computerized brain transmissions are revealing pictures of human behavior and *uber-beliefs*.

It is of little wonder that so many are anxious to '*join up or pay up*' for the promise of some extraordinary *spiritual experience*. The awesome wonder of mystical, transcendent experiences—wrought by group meditation, group drumming, prayerful rocking and reading, near death experiences or a hundred other activities—are frequently defined as *supernatural* by pious leaders and have been for thousands and thousands of years. Until Newberg did his progressive SPECT scan research and demonstrated otherwise, the explanations for such *oneness with the unseen universe* had been ascribed to a myriad of deities: Allah, Brahma, Jehovah, forest tree spirits, pan-theism or even demons such as Satan or Beelzebub.[10] Now we can know that these alleged supernatural epiphanies are a *to-be-expected function* of natural and normal brain physiology. Here you may suggest that *'God'* placed that experience within our reach and therefore it is no less valid; since the ultimate objective is a *sense of oneness,* the explanation doesn't matter. But once again the more rational person will appeal to Occam's razor (the Law of Parsimony) rather than to mystical or metaphysical rationalizations. The simplest explanation, Bishop Occam suggested, is consistently closer to legitimacy. Complex, complicated, convoluted explanations involving deities or spirits from other realms has in modern information-oriented times become the hallmark of the uninformed or the ignominious.

Spirituality, for example, does not require some convoluted contorted explanation. It is a naturalistic vision of the universe and us. To have *spiritual* experiences can be readily understood as psycho-neurological states constituted by the activity of our brains. For example, when exposed to astounding stimuli and parts of the cortex 'light up' it is reported as awesome. When one is in a state of awe and wonder it can certainly be defined as 'spiritual'. This in no way lessens the attraction of such experiences, or renders them less weighty. Appreciating the natural physical world and our complete inclusion therein engenders feelings of union with all that is. The truest definition of *Spirituality* is not complex, stodgy nor sticky sweet but is rather found in the context of *awe* and child-like *wonder* about the grandeur of life.

Oliver Sacks in his fascinating book *Musicophilia*[12] turns a neurological narrative about a brain-damaged patient into a palpable appreciation of *spirituality*. One of his patients had a sudden rupture of a brain aneurysm that bled out into the frontal lobes. This *accident* resulted in this bright MIT graduate losing his *wonder*. He said: "*Wonder* had been at the core of my previous life". It seems that a long with future planning and *mystical one-with-all experiences*, w*onderment* too (*spirituality*), is a function of frontal lobe activity.

While some leaders may chose to define it otherwise, we now know that mind-altering spiritual experiences, (*wondrous*) described through the centuries as '*holy being' inspired*, are clearly the product of neurological synaptic firings. While it may be argued that Deities use these same avenues, that kind of thinking has led more than one cult down the Kool-Aid path. TMI?

Announcing and then clinging to ancient answers that can be neither confirmed nor denied has been the forte of self-styled seers and con artists for eons. Such fixations stifle cognitive change and development. We may smile knowingly at the old story about the ancient oracle that was alleged to have said "a great army will be defeated today." The ambiguous pronouncement turned out to be absolutely true! We may express amusement but today we still hear sanctimonious leaders or *'spirit channelers'* rhapsodically lay claim to ambiguous sayings and choose to define such experiences as *spirit inspired* and therefore *spirituality* in and of itself. Once again, random correlations become divine causation. However, as time passes such definitions only provide more material for stand-up comedians such as Penn & Teller.[13] It seems self-evident that to survive modern-day silliness we must cease our chronic kvetching about certain mysterious correlations and begin renouncing the *dupers* who eloquently provide ambiguous, comfort-creating explanations. This will not happen via some mass movement. Only as the single individual sees the absurdity of our now almost universal *silliness*—often called *sacred*—and begins to LOL, or at least smile, can change be accomplished. It is the heroic insightful individual who will stand in the face of a majority or a charismatic diviner and call out the non-sense. When you hear, "have faith" or "that is a religious issue and therefore we cannot call it silliness," that is when you start to laugh, yes, even out loud.

Your willingness to laugh at yourself can become an epicenter of rational thinking and continued physical brain development. As a beloved friend once said, "If we can laugh at and about our absurdities years from now—why wait? Why not begin laughing now?" Such laughter is, after all, the beginning of understanding *the cosmic joke*.

CHAPTER 4

DUPED BY DESIGN

Evolution vs. chance

In many a *nodulized* mind-set, evolution has been a provocative topic for as far back as their memory can take them. I first heard about the theory in middle school science class. I sat in the front row and developed a heart-wrenching crush on my teacher. She had an unbelievable tan and wore skirts that accentuated her bronzed legs—crossed and uncrossed frequently. At that same time BION (believe it or not) I did learn! I learned the vital key for a theory to be valid: it must be "able to predict and have those predictions repeatedly replicated by objective experimentation." Yes, I had some great fantasies going on, but no reality based, empirically justified, predictions.

Today, most scientists acknowledge that the Theory of Evolution has astounding predictive power. However, this kind of fact tends to send some anti-science people into apoplectic states. The internationally published decision by a 2004 school board in Pennsylvania to treat alternative explanations such as Intelligent Design (biblical creationism) on par with evolution created a brouhaha heard 'round the world.[1] (i.e. Proposition that life's complexities could not have happened without the intervention of a supreme intelligent force. e.g. Allah/Jehovah or an omniscient watch maker.)

A federal judge heard the case and posed the questions: Are Intelligent Design and/or several other alternative explanations of creation creditable scientific theories and therefore eligible for tax money? That is, do they deserve to be taught in public school science classes?

The plaintiffs made the case that Intelligent Design does not qualify as a scientific theory at all. As mentioned above, <u>predictability</u> was central to the whole case. Simply stated, "evolution theory allows for predictions that can be empirically validated." The other side argued from Aristotle (This line of reasoning was covered in chapter 2.) that the seemingly logical events found in nature, including the ultimate earthling Homo sapiens, could not be random happenings but rather must be ordained proceedings created by a greater omniscient intelligence with a specific design plan—in short: the work of a God. And therein lies still another major rub: which God?

When the Founding Fathers established the Constitution of the United States of America they went to great lengths to include what some have called "excessive documentation" related to separation of church and State. They did their utmost to be sure there would be no despoiling of what could become the grandest experiment ever in forming a government of the people, by the people, for the people—not just the non-believers—not just the believers—but all people. The Founders were particularly careful to ensure the right of minorities. No tax or state sanctioned preferential treatment would be given to any religious group or teaching. Clear as a bell! Actually, separation is also about protecting religion. Notwithstanding the remarkable secularism of most of the leading Founding Fathers, their aim was the liberation of religion from government as much as the reverse. Furthermore, when you stop and think about it, the disestablishment of religion is actually somewhat responsible for the religiosity evidenced in America.[2]

Notwithstanding the vociferous objections of some anti-evolutionists, for over a century now, objective scientists on every continent have been tearing into this planet's crust and unearthing evolutionary evidence. Others have been working from within the molecular composition of DNA and they found evidence that shows, without a doubt, that the principles of evolution are in fact—fact. We now know with absolute certainty that every organism on the face of this globe is subject to the Laws of Evolution.

QUESTIONS KEEP COMING

While some things in our universe remain an awesome mystery for the present, answers are continuously forthcoming because scientists keep asking questions! Furthermore, most rational researchers recognize that as one question is answered, another arises. This is implicit in the scientific research endeavor. In fact, the search to answer so-called unanswerable questions is the adrenaline that keeps most researchers going. For some others, such open-endedness is abhorrent. And therein, we discover the real discord between Creationists and Scientists. Some human beings seem to relish and celebrate "not-knowing-and-searching", while others seem to need an absolute be/all and end/all answer and if not a factual answer they'll accept a fable, as long as it fits their preconceived ideas (*nodules*) and provides comfort.

The question that faced Pennsylvania's Dover School Board was whether or not the imposition of one particular creation belief (Judeo-Christian) on a multi-ethnic, tax supported, secular student body is in keeping with the law that prohibits the creation of a state religion. If they allow one belief system to be taught, surely they must also use tax dollars to teach other belief systems? Still, says the New Scientist, about 16 percent of high school biology teachers in the US are Christian Creationists and half of them, despite the court-ordered ban on teaching what is called *Intelligent Design* in science classes, publically declare they do instruct students regarding Creationism as written in Genesis as a valid alternative to Darwin's Theory.[3]

Let's regroup: A theory, to be considered viable, must have empirically based, objective data supporting it. Specific predictions must be replicated repeatedly. For example: Recently the number of harmful mutations in chimpanzee DNA was predicted by knowing the number of mutations in the Homo sapiens DNA. Who would'a thunk it? Yes, not only Darwinian theory, but all theories must be experimentally validated and the supporting data become facts by predictions that are <u>repeatedly confirmed</u> by empirical scientific evidence. Just GS.

On another and still larger front, the basic tenet of the entire scientific method is the ability to construct a set of circumstances in which a statement (hypothesis) is made and then an outcome predicted and finally consensually observed. Consider for a moment how we can predict a solar eclipse based on the theories of Newton and Galileo. Today we accept as fact an Email or TV news announcement of an upcoming eclipse without a second thought. We do not <u>hope</u> the announcement is correct, we know it, because when any of us arrive at the predetermined place, at the announced time, we can all see the eclipse with our own eyes. Thus the theory is consensually validated. There is no "we believe", "we hope" or "just have faith". No, it happens just as the theory said it would!

Another very important ingredient in the validation of a theory is the ability to empirically test a non-obvious theoretical precept. For example, the Theory of Evolution should be able to predict that a harmful mutation would die out if it were serious enough to reduce the odds of leaving descendants. That precept proved true for mice, chimps, and dogs.[4] Evolution's constant power to predict the unexpected is the super-glue that allows the morphing of its datum into 'facts'. This is particularly true when the prediction is repeated and validated in all parts of the world—time and time again. Charlie Darwin also hypothesized that the biochemical basis of emotions utilized in various survival ploys would be highly conserved in the process of evolution. And sure enough, Candice Pert gave credit to Blanche O'Neil[5] for finding the same kind of neuro-peptides that mediate human emotion in one-celled animals. One-celled critters survive just as we survive using the same basic molecular chemicals to facilitate a quick get-away.

Today, many fundamental religionists are a little anxious because try as they may, they cannot invalidate the plethora of scientific data flooding out of international laboratories. You will remember that invalidation of scientific fact was unsuccessfully tried by the religionists of Galileo's day. It is equally important to remember that neither Copernicus nor Galileo were particularly beloved by the popular majority for informing us that the Earth was not the center of the universe. Galileo, in particular, did not endear himself to the believing masses by saying things like "I do not feel obliged to believe that the same God who has endowed us with sense, reason, and intellect has intended us to forgo their use." There are still millions of Americans who resent Charlie Darwin for putting all of us on the same tree on which, bonobos, chimpanzees, and gorillas also have nearby branches. That's just the way it really is! TMI?

Now, at this very moment in human history, after a miraculous landing of NASA's new spacecraft on Mars, we may discover the skeletal remains of microbes that once flourished beneath Martian soil or if not there, certainly on one of the other planets recently discovered. Please allow me a small but exciting segues here.[6]

Astronomers have recently announced (Fall 2010) that they have spotted a planet in what is called the Goldilocks zone: Not too hot, not too cold. Co-discoverer R. Paul Butler of the Carnegie Institution of Washington said, "it is Juuuust right. Similar to Earth in places!" This new planet is unlike any of the nearly 500 other planets that astronomers have found outside our solar system. Life on other planets could be entirely different. Life doesn't mean E.T. or even the equivalent of shower mold. Temperatures in the new planet are estimated to be as hot as 160 degrees or as frigid as 25 degrees below zero, but in between—in the land of constant sunrise—it would be '*shirt-sleeve weather.*' "Furthermore, chances for life on that planet are 100 percent!" said co-discoverer Steven Vogt of the University of California at Santa Cruz.

Meanwhile back on Mars: Researchers are also elated because data received back from NASA's *Phoenix Mars Lander* suggests liquid water has interacted with the Martian surface throughout the planet's history and into modern times. The *Lander's* robotic arm dug into the permafrost and deposited samples into small ovens, which were then checked for the base building blocks of life—water vapor and carbon compounds. Indeed, life does find a way, whether in the underwater sulfurous, superheated toxic volcanic vents of Hawaii (where life thrives on superheated toxins as if in another distant planetary environment)—life yearns for life! Actually, there already is an alien life on earth. BMA! Consider this: NASA microbiologists 'taught' bacteria to live without basic 'life biochemistry'. Over a few months the bacteria learned to replace our Earths 'life essentials' with things like arsenic. These bacteria thus became a life form never before seen on this planet. NASA has changed the definition of life and how we look for it on other planets.

It does seem logical that in our ever-expanding universe with more than 100 billion galaxies, each containing an average of a hundred billion stars, the discovery of alternate life forms is only a matter of time. As expected, Vatican scientists have already reconciled this new scientific revolution with their archaic doctrines by announcing that the discovery of extraterrestrial brothers and sisters would not contradict Christian teaching.[7] Whoa! Most American religious fundamentalists would say that the Pope is once again over-reaching. But when the discovery of an ET approximation is made, the Pope's Astronomer will no doubt be elevated to sainthood. J/J (just joking)

WHO WANTS YOUR VOTE?

Even more astonishing to scientists and to some in the political arena, is the fact that during the last major election go-round, popular presidential candidates were careening from blue state to red state reproachful of Evolution and propping up the *uber-belief* in the creation story as it is explicated in the first pages of Genesis. Hopefully the 48 percent of Americans, who, up to this moment in time, apparently still cling to the literal Genesis story, will begin to treat new scientific data with some modicum of rationality. While the Church leaders of Galileo's time attempted to undermine the newly revealed science and provided an epic drama—that only recently resulted in an official ecclesiastic apology. (Obviously, Vatican VIPs don't want to repeat that Bruno/Galileo fiasco again.) Interestingly, as is true today, thinking people of Galileo's day just shrugged and incorporated the new data into their lives. One of our more provocative modern philosopher/scientists has this to say: ". . . the idea that our planet was not the center of creation . . . sat rather lightly in people's minds . . . Every school child today accepts this as the matter of fact it is, without tears of terror."[8] Most fundamentalists in Galileo's day—even after the uninformed bullying and revered proclamations made by their church leaders—were able to accept and incorporate the new geo-science into their sacrosanct systems. Unfortunately, even while this incorporation occurred, Galileo was singled out and spent a lifetime under arrest.

While a segment of American population continues to hold to "pre-Galileo" beliefs American politicians will exploit them with rhetoric no matter how inane.

DENIAL SURVIVES AND NODULES GROW

On one hand, it seems that for a vast number of Hominids inhabiting this planet, a case must be made that only the Almighty can create life. On the other hand, perhaps it is time to announce and espouse a larger-than-ordinary expression of creation and its creator. The ultimate question for Galileo was not whether or not the Almighty was capable of creation. His question had more to do with the prevailing science of his day. The petty definitions of the Almighty that religious leaders often announced as ultimate truth were not important to Galileo. Following his lead, perhaps a more significant approach today might be: "How can we incorporate the new ever evolving scientific data and expand our limiting beliefs about the "Almighty Creator"? Recent neuro research could free us from having to defend medieval *uber-beliefs*. Die-hard believers could now be acquitted on the charge of worshipping Divine Personages who possess a Homo-sapiens temperament and characteristics. More importantly, we no longer need the impossible word Omniscience to explain the unknown. We see mysteries unfolding to our view daily. The incomprehensible is becoming comprehensible!

CREATION RESEARCH

Now that the first complete DNA for a one-celled organism has been built from the ground up by genetic scientists in Maryland, questions of ethics and even morality will no doubt spring from the confused but popular politic. Building an entirely man-made chromosome was supposed to be a role reserved for an omniscient intelligence. So this new laboratory creation is especially angst provoking to some, since non-believers created it out of inanimate proteins. In the next few months, these twenty-first century alleged "Doctor Frankensteins" will transplant the creation into a microbe and it will just boot up like a Bill Gates' software program. Certainly the nervous giggle accompanying their work will resurrect memories of the cult classic movie where Dr. Frankenstein (Gene Wilder) shouts: "It's alive! It's alive!" But this modern drama is not a spoof. Unlike the Sci-fi movies—the slick laboratory sets, the high tech props, the bubbling beakers, and white coats containing obsessed scientists—everything in this spectacular scene is as real as it gets!

Frightening as it may seem, or actually be, human beings are usurping the singular hallmark of the classic medieval 'Creator-Deity'. All around the world, researchers are creating life out of inert matter. Given that according to most fundamental religionists only the Almighty should be able to bring forth life from inert dust or ribs; this 'creative activity', must smack of sacrilege for nearly half of the American population. At the same time, excited genetic scientists in laboratories in almost every major country are replicating this astonishing work. They are creating life from lifeless materials much as the way the Judeo/Christian Bible describes Jehovah creating Adam out of a handful of moist mud. To the chagrin of some, this kind of creative activity is becoming commonplace.

DEATH AND RESURRECTION RESEARCH

When dedicated *Genome Geeks* on both sides of the Atlantic first completely mapped the human Genome, they were met with two rather quaint surprises: First: There were only about 22,000+- genes found in the human genome. This was probably a little humbling since they had expected upwards of a hundred thousand. Even the lowly water flea has 31,000+- genes and while some are new to researchers, it shares many others with Homo sapiens making it a surrogate model for researching toxin impact on human beings.

Secondly: Our human Genome is littered with bits and pieces of material, which did not appear to be purposeful—so offhandedly they called it 'junk'. It turns out that a lot of this supposed 'junk' is an ancient viral graveyard. It is a microbiologist's dream—overlapping the role of a micro-paleontologist—to unearth thousands of viral fragments, buried for centuries like ancient dinosaur bones in the Pre-historic Los Angeles La Brea tar pits. Yes, startling as it may seem at first, viruses long extinct left their 'fossil bones' behind in this modified hominid genomic 'tar pit', euphemistically named 'junk'.

Scientists in Parisian laboratories marching to a slightly different fife and drum combined the tools of genomics, virology and evolutionary biology. They pieced together the broken parts of one of the microscopic viruses (junk) that had been dead for thousands of years. Yes, since it may be hard to believe, I will say it again: human DNA is literally littered with the residual remnants of ancient virus fossils! In another "Dr. Frankenstein-like" maneuver, French molecular geneticists at the Gustave-Roussy Institute[9] actually reassembled some of these long-dead viruses into their original living form. They brought the ancient viruses back to a living form and carefully placed them into human cells. (Cells that had been removed from their human donors.) The newly resurrected creations inserted themselves into the DNA of those human cells and began to thrive.

Apparently viral vestiges, scattered throughout our genome, have long ago lost their infectivity due to mutation. By comparing ten or so of these dysfunctional fossils, Thierry Heidmann and his colleagues puzzled out the components of the original virus that infected our human ancestors eons ago. The reconstructed version, used to infect the Petri dish human cells, was called *Phoenix* for obvious reasons.

Heidmann also claimed that his work was completely harmless and argued that since resurrected viruses could play a role in cancer research—because high levels of similar viral particles have been found in many cancerous tumors—such research should be continued.

Others, like Rutgers University molecular biologist Richard Ebright,[10] are more critical of virus-resurrection experimentation and say that while this kind of research isn't likely to yield medical applications or bio-weapons, such work should be subject to international oversight. To cap off his objections he cautioned that until the virus was resurrected, its infectivity could not be unequivocally predicted. For many, that warning seems scary enough to curtail research.[11] Scarier still, is the fact that there are no U.S. laws guarding against the laboratory creation of human pathogens other than smallpox. In fact, poliovirus, the 1918 flu virus, and now this previously extinct virus have all been assembled in the laboratory.

Many scientists agree on the need for some form of oversight. Other, perhaps more audacious researchers understand that creating life or bringing something dead back to life conjures images of Spielberg's *Jurassic Park* drama or better still, the story of Jesus raising Lazarus from the dead. These same scientists are much more interested in how such life giving experiments can provide clues about how HIV or other even more devastating viruses work.

Other aspects of the 'Junk' genetic material appear to be vitally important in turning other genes and gene complexes on and off. So once again, is the 'junk' of one generation the treasure of the next? IGTP

Robert Milton Ph.D.

WHEN IS A THEORY NOT A THEORY?

When genomic data demonstrated the exact order of all 3 billion bits of genetic code that go into making Homo sapiens as well as bonobos and chimpanzees, few of the researchers were surprised that the genome-sequences of the three were more than 96 percent identical. Charlie Darwin deduced a similar result in the mid 1800s.

Unraveling and sequencing the chimpanzees' DNA allowed more than just a confirmation of how similar humans and chimps are—it put the Theory of Evolution to a tough test. If Darwin made the correct guess in the 19th century with little more than a visit to the Galapagos to support him, then modern genomic geeks in the 21st century should be able to make specific predictions about the number of harmful mutations in chimpanzee DNA by knowing the number of mutations in homo-sapiens's DNA.

Darwinian theory makes sense only if humans share most of the viral fossils with what Charlie referred to as common ancestors—that is, chimpanzee relatives. Sure enough, when Eric Lander, a geneticist at the Broad Institute of MIT and Harvard in Cambridge, Massachusetts and his group[12] computed the risky mutations in the chimp genome, the number was a perfect fit to the prediction. These same researchers calculated that if this finding were a mere happenstance, chimps and humans would have had to endure a googolplex number (largest integer conceivable) of identical virus attacks in the course of thousands and thousands of years; then by some unknown roll of the cosmic dice those infections would have had to end up in the precise same place within both the chimp and human genome. This is as about as probable as building a Boeing 747 out of the scrap in desert junkyard filled with rusting freeway wrecks. Seems I've heard that analogous argument somewhere before. (?)

The helical rungs of the human DNA ladder include something over three billion pairs of nucleotides filling appropriate places on the forty-six chromosomes. The unique sequence of these nucleotides precisely determines how each and every human being differs from every other human being as well as from every other living creature. Logic tells us that only way a human being's genome could possess the exact viral DNA found in another species (e.g. chimpanzee) is by becoming an heir to it. (i.e. via a common ancestor) Today, we have meticulous comprehension regarding that inheritance. Enormous databases of genomic sequencing make the recreating of ancestral genomes almost commonplace, so much so that you and I, for a not-so-nominal price, can obtain a genetic record of our ancestry. Genealogy ain't what it used to be!

As Rick Weiss of the Washington Post said:

> "Their analysis was just the latest of many in such disparate fields as genetics, biochemistry,

geology and paleontology that in recent years have added new credence to the central tenet of evolutionary theory: That a smidgeon of cells 3.5 billion years ago could—through mechanisms no more extraordinary than random mutation and natural selection—give rise to the astonishing tapestry of biological diversity that today thrives on Earth."

The Theory of Evolution has repeatedly passed all the tests. Like the Copernicus' Heliocentric theory, the vast majority of reputable scientists, now considers Darwin's Theory of Evolution, factual.

INCORPORATING FACT INTO DAILY LIFE

Since the days of Galileo the plethora of scientific facts have often threatened established *uber-beliefs* and ecclesiastic domains. Instead of brushing the new data aside and threatening to punish today's *Galileos* (by terminating their funding thus terminating their research), modern believers could and should seek ways to integrate the new discoveries. Expanding our concept of the "I am that I am" to include black holes, dark matter, expanding universes and myriads of inhabitable planets is a small step for thinking men and women but for some others it seems to be an insurmountable *nodulized* profanation.

What we popularly call the average human mind has been proven to be a biased, *nodulized,* spin-doctor not an objective CEO. Advanced neurology has demonstrated that our human conscious mind/brain, rather than eagerly grasping new innovative research, is more likely to merely offer rationales and easy stories to comfort and reduce anxiety if the new research produces stress. Furthermore, recent genetic research has shown that we may have no more free will choice about certain aspects of our behavior than we do about the color of our eyes.[13]

But, instead of ranting about how science is destroying our belief in our concept of a Divine Personage, maybe we should begin to expand our concept of the "Divine". Perhaps, when the inexplicable occurs, children may have the right to fantasize about explanations. Fairy tales, including the Wizard of Oz, and the Easter Bunny may play an important part in a child's cognitive development but for thinking adults to jump to fanciful conclusions is just nutzy. For example, there is a joke about a small town hardware store owner who gained a reputation for miraculous cures. A little old lady all bent up, slowly shuffled into his store and within ten minutes came out walking completely erect with her head held high. A customer notices and shouts, "It's a miracle! You walked bent in half and now you are walking erect. What miraculous thing did he do?" She answered, "He gave me a longer cane."

Published, peer reviewed, scientific data probably says as much about our wonderful evolving Nation at large as it does about specific scientific truths. Paradoxically, America appears to be simultaneously post evolutionary and pre-Darwinian. We are willing to spend billions on war materials and most kinds of research and yet grimace with disbelief when some of the recorded data is laid bare. We claim we are desirous of change and then scowl and hiss in unison with pious leaders who say "that is dangerous science". Agreement on current social issues—abortion, global warming, gay rights, etc.—is not likely in today's climate of Right values vs. Left liberty. Once again the nodules rule! Yet, if we cannot reach consensus or at the very least hold a rational dialogue on where we came from and how we survived in the past, how can we expect to agree on where to go and how we should survive in the future?

CHAPTER 5

WHEN IS NEW REALLY NEW?

"Less is more." Occam's razor

Almost every day we are startled by news bulletins associated with what is called "our swiftly changing world." Most agree that our world is morphing—but how and into what? These are the questions. Trying to balance between the extremes of crippling cynicism and giggly optimism has resulted in a whole myriad of wanna-be political and religious prognosticators, all of whom have alleged *valid* predictions and so-called *original*, innovative theories, which for the most part, merely mask their own personal desires and disabilities. Does this sound cynical? SWEIN?

It does appear that technology has provided something new by flattening and shrinking our world beyond the recognition of those now collecting Social Security. The field of communication alone has changed more in the last twenty or thirty years than in the previous three thousand years—or even three million. But is it the world or its inhabitants that are changing? Or both? We know that Gutenberg's press changed the way we moderns think: i.e. logically, linear, disciplined and in more complex metaphors. In a similar manner the universality of cell phone *texting* and, if its incumbent acronym craze continues, we will no doubt change our thinking to an abbreviated format of minimal informality. Subjectively we (our brain) morph to accommodate the new technologies. Not only does a repetition of thought mollify the brain, repetition can change it, grow it and rewire it. That is why it is so important to learn to '*load content*' that is conducive to upbeat character and constructive human interaction.

AGE OF ENLIGHTENMENT

The Age of the Renaissance was thought to be the beginning of worldwide transformation primarily because of the sudden proliferation and spread of scientific and technological information. The sea voyages of discovery in the 15th and 16th centuries were like large-scale explosions spreading *nodulized* European thought all over the globe. Instantaneously, so it seemed, the world began to shrink. But take another look at the historical facts: Thinkers, philosophers, scientists and missionaries began to *exchange information*. Biased information, no doubt, but information just the same. In a matter of months they could investigate a section of the globe, exchange ideas and synergistically expand the indigenous people's *nodules* with respect to European beliefs and folklore. By comparison to previous centuries, Western civilization's ideas were merging with other civilizations at an incredible pace. Undoubtedly, brain *nodule* growth was keeping pace on the new continents. In every corner of their newly accessible world, zealous missionaries found that they could easily win converts. They began to spread their *nodulized* point of view and since it was attached to relatively advanced technology, which undoubtedly seemed like magic to indigenous peoples, the *new religious ideas* were eagerly accepted and practiced. We've already discussed what repeated *practice* does to the plastic human brain in chapter one.

Today you can find a seat in almost any Starbucks coffee shop in America, Europe or China and access the whole of *nodulized* human experience on your laptop computer—at least most of it. But today's worldwide transformation is more like an all-encompassing information *implosion* than an explosion. Motherland patriotism and the need for national distinctiveness seem to be falling fast asleep in most progressive countries. What we usually think of as *highly diverse countries* are folding up their walls and becoming commercially unified and borderless. Europe, for example, which has traditionally been a mish-mash of nationalistic chaos, has set a stiff pace for unity, commercialism, borderless-ness and an anti-war mind set.[1] But once again we must ask: Is it the *world* or is it human ideology that is imploding? Today's missionaries for instance, are finding it difficult to win great numbers of converts with their traditional *nodulized* rhetoric except in poorer countries where superstitions are still an intrinsic part of daily life and where the *Internet* has not made dramatic inroads.

History and this world, as we have perceived them for thousands of years, have come to an end. You and I are no longer inhabitants of a state or country with a fence or ramparts to fend off foreign insurgents. Even though we may consider ourselves defined by our patriotism or chauvinistic pride, we are—like it or not—*world citizens*. In part, we're hearing the last gasps of nationalistic flag-waving in the form of attempts on the part of China and India to become capitalistic and at the same time remain independent of each other and the US. The familiar history, involving singular territorial segments, unique monetary systems and intricate language differences are also gasping for breath and a new age has begun wherein we are all citizens of a planetary civilization—and many of the *'less developed'* countries are lovin' it!

THE REAL NEW AGE

Another problem arises, when we analyze this newly developing mind-set and realize that there are still large population segments in every nation that have grown up being taught to *believe* uncompromisingly rather than *think* logically. So while credible scientific information was and is being published on a daily basis—divergent from long-established *nodulized* beliefs—many *uber-believers* choose to discount it. For instance, when the words 'New Age' came on the scene three or four decades ago many ardent *believers* quickly adopted, modified and incorporated the purported *new data* to fit their old ideas. Since new science paradigms such as Quantum Physics were/are barely understood by the general population, a few modern meta-physicists began to adapt certain aspects of the new age sciences and integrated them into so-called *'New Age'* spiritual systems. The old belief-*nodulized* doctrines merged and emerged with a new vocabulary. Just like the old authoritarian missionaries, the popular *'New Age'* missionaries blended science with fable to impress the uninitiated.

Actually, if you look closely, much of *'New Age'* literature is a reincarnation of old beliefs. Titles of books reveal the recasting: *Out of Body, 90 minutes in Heaven, Is There Life After death? UFO Abduction, The Return of the Gods, Astrology, Spiritualism, ESP, Tarot, Psycho-kinetics, The Secret.* These and hundreds of similar titles are available to the recently emerged New Age *believer*. Most of these books are just the former large-*nodule uber-belief* systems wrapped up in new scientific sounding jargon. These kinds of books, according to publishing demographers, sell at about the same rate as pornography. In fact, the philosophical novelist, Jostein Gaarder[2] says that the difference between real philosophy and these kinds of books is roughly "the same as the difference between real love and pornography." Ouch!

Obviously, it would not be fair to say that all of these kinds of writing—Parapsychology, ESP, clairvoyance, cosmic consciousness, spiritualism, UFO-ology, astrology—are equally bogus but these themes, especially when taken all together, explicate a realm that does not exist except in the mind/brains of *nodulized* believers. It is another fringe-science nostrum that cannot be proved or disproved—only believed or disbelieved. Most of these books have no truck with the actual scientific method and are not open to serious objective investigation.[3] In both England and America there are large rewards posted for any person who can offer the slightest proof of something *extra* sensory or *super* natural. The reward has yet to be collected. (See: SkepticsSociety.com)

Like pornography, the publication of such books has become an industry in and of itself. A billion-dollar business has sprouted by linking statistics to so-called supernatural miracles. As everyone knows there are lies, damn lies and statistics. The published efforts to find causation (fact) within correlation (statistic) has created a new bevy of converts who are, in fact, hooked on *believing* but at the same time wish to link their *belief* with legitimate *controlled method* research and then bestow a scientific name on it. For instance, the published discoveries of Quantum physicists are borrowed and then interpreted to mean that the universe is in fact, literally *one giant quantum field* in which everyone (and everything) is inter-connected and can influence one another directly and instantly. Quantum Physics is also directly applied to the definition of *consciousness*[4] so that the brain becomes a translation device whereby the larger *quantum world* is represented through electro-biochemical signals. Such jargon may sound impressive but utilizing scientific nomenclature in these explanations, most of the time, reveals two systems that are by and large, incompatible. Sort of like trying to weigh a *soul* (see End note 3) with a laboratory balance beam scale or like confusing kangaroos and wombats.

While visiting the giant redwoods in Northern California I found myself surrounded by a throng of college students. They talked and walked around the trees, hugging them and calling out various times, "It's 4:10 for me. Oh, it's way past 4:20," another would chime in. Finally, I asked what the time references were about and was told that it referred to the "actual clock time high school kids met to get stoned". In that part of

Californian 4:20 was the time they met after school to smoke Marijuana, so various times around 4 o'clock took on the meaning of subjective levels of being high. While I was a bit confused, none of them considered that they were talking about clock-time per se. When physicists talk about "quantum this" or "quanta that," they are talking conceptual abstractions that have no actual connection to everyday life as we know it—or to the descriptions used by 'New Agers'. When writers pick up on quantum jargon and apply it to their special belief systems it is like insisting that it is 4:20 in the afternoon at the morning AM breakfast table. It makes no real sense—and has no real connection to time—except to a limited few who happened to be 'in' on the *stoned* meaning of 4:20. (However, I just realized you could be smoking pot at your breakfast table—it may make sense to you and Deepak—sorta.)

INVENTION VS CREATION

Fifty years ago a sociologist named Robert Merton[5] asked a very exciting question about *genius* and scientific discovery or *inventiveness*. He found that the same (or very similar) inventions often appeared in a parallel time frame—in other words, the same invention seemed to 'pop from the minds of geniuses' in several parts of the world at about the same time. He concluded that genius is not a unique source of insight but rather an efficient source of insight. But his observation of scientific genius proved **not** to be the case with '*artistic Genius*'.

For example, Mozart was considered a musical genius while his celebrated teacher Salieri was not. The work of the original artistic genius is singular and unique and whole groups of individual geniuses may flock to specific '*schools*' to add their uniquely creative strokes but it is the mentor who is recognized. Our persistent inability to separate scientific genius and artistic genius can be troublesome. Mozart's music became more popular as years passed. Salieri's music was largely forgotten or intentionally ignored. So we can easily conclude that if we combine the talents of a dozen Salieris we would not arrive at even one *Magic Flute* opera. Another example might be allowing a team of bright, talented art students to be placed in a French riverside park every Sunday for two years. To be sure, this kind of experiment would still not result in an original George Seurat *Sunday in the Park* pointillist portrait.

Shakespeare created Romeo and Juliet. He fashioned them and gave them form and life. He '*owned*' them as no one before or since could. History says that Alexander Graham Bell '*owned*' the telephone, but not as well known is the fact that Bell's ownership was only because his patent application was accepted a few hours earlier than a competitor's.[6] Shakespeare's creative genius is of a unique variety while what, most of the time, we call *inventive genius* more likely the product of the *zeitgeist* and *think tanks* across the world. Thus a multitude of similar inventions pour forth in precise unison. These inventive whiz kids, Merton said, "march to a synchronous beat while the artistic genius marches to a distinctively unique beat."

Notwithstanding, there is a great deal of confusion precisely because the same or similar words are used to describe both kinds of genius *inventors*. It is this same kind of confusion that arises when a few well meaning *belief specialists* (meta-physicists) attempt to impose a paradigm of physics on a system of *ideas* where it doesn't belong.[7] This, JMO, is an important concept and will be repeated in different ways throughout this book.

Even while a new world-wide paradigm is emerging right in front of our eyes in terms of neuro-imaging and genetics, many '*nodulized* believers' insist on the old *inspired revelation paradigm* or '*new age metaphysics*'. Basically, their hue and cry resembles the age old, "there just has to be something more out there. Three score and ten is just not long enough! Someone or some '*it*' out there just has to solve my quandary." They also frequently insist that there must be *special people* who have *special powers* to reveal *special insights*. Sounds very much like the old 'oracle syndrome'[8] wrapped up in new terms like *channeling*, universal *quantum consciousness, and web of resonation, Boehm vibes and* etceteras.

OTHER ALTERNATIVES

Most likely it was curiosity and the desire to understand the great mysteries that propelled early hominid troglodytes out of their caves. Their construction of explanations via pantheistic mythology and comforting, *anxiety-reducing bedtime stories* is perfectly understandable. The need for charismatic chiefs to lend authority to such explanations is also understandable. In a previous book I went to great length to describe this process. Debate has continued ever since the cave exit and perhaps has even grown more intense since the emergence of New Age ideology characterized by primitive rococo flourishes. The attempted amalgamation, for example, of Quantum Physics with ancient Curandero / Shaman *out of body practices* has created another kind of folklore, which slakes the curiosity thirst of many, simply because it provides an un-testable thesis.

By most definitions[10] a *Shaman* is a specialist and master of the ecstatic trance-journey, only in 'New Age' parlance has the *Shaman* become a synonym for tribal healer, holy person or medicine man. Traditionally, bona fide *Shamans* practiced ecstatic trance-journeys, and developed the typical beliefs and healing techniques that arise from what are usually call psychedelic journeys. In short, the authentic traditional *Shaman* is said to communicate and find answers in the *spirit world*. In this system, Spirits can be good <u>or</u> evil. (A revisit of dualism) Local *Shaman* healers (curanderos) most often use psychedelic plants as a way of '*seeing other dimensions*'. Some modern *believers* still seek the Shaman, Fortuneteller, and intuitive *Readers of tealeaves*, sacred writings or icons to provide something mystical or transcendental. But *nodule* addictions of this type, when repeated, become fixations that typically will not allow scientific views to be entertained.

CORRELATION AND CAUSATION

When we think of the myriad of things that can happen in a single 24-hour segment of time, it is astounding. Take a moment right now and think of all the occurrences and all the different things you experienced in the last 24-hour period. When you focus, more often than not, you realize, strange coincidences do happen all the time. E.g. your phone rang just as your automatic garage door opened without pushing the remote. Your sprinklers came on without the timer being manually activated and at that precise moment the grandfather clock chimed at an inappropriate hour on Friday the thirteenth, which happened to also be the night of the blue moon. Ad infinitum. GMAB!

The unspoken point of much of the above-mentioned gobbledygook is that human beings love, seek out, and embellish coincidences. Do we love them? Hey! We collect them, bind them in books, revere them and then make religious rituals from them.

Of course, it seems obvious that some brilliant people are *hard wired* to latch on to strange, seemingly inexplicable experiences, remember them and even write about them! Understandably some of these bright articulate individuals shift their emphasizes to *nether-world interpretations*. Since their *nodules* impel them to perceive what they have practiced in the past, it really does seem *otherworldly*.

Most primitive peoples believe in some kind of magic and sometimes even survive by connecting coincidence upon coincidence. If an animal appears at a certain place at a certain hour and the primitive hunter notices, it could mean the difference between nourishment or starvation, life or death. This kind of connection has a certain kind of AWE-some-ness to it. When such data is taken from the coincidental experiences of millions of people and cobbled together into books it begins to look like genuine supernormal prediction. VOILA! Correlation becomes causation!!

Average, normal people forget—actually, we are expected to forget and we do. Politicians and prophetic prognosticators depend on it. But bright, intuitive, insightful people *remember*—sometimes without their conscious awareness. They remember details that the rest of us were hardly aware of in the first place. It was Kundera[10] who suggested that the struggle of man against power is the struggle of memory against forgetting. Yes, those who have unique memory capacity also have unique powers and they know, all too well, that most of us forget. But they are not supernatural nor do they respond as the result of some other-world-spirit connection; rather, it is our *Pollyanna* brain in action that makes it appear so.

Even before Dale Carnegie explicated the subtitles of salesmanship, it was known that smiles attract—frowns repel. Recent research has revealed some other super body-language subtleties. The ability to influence others non-verbally has resulted in a fascinating compendium of psychological studies.[11] They show that what some others have called the *connectedness of cosmic consciousness* can be 'objectively' studied via non-verbal communication. Again, the 'duping' action of the *Pollyanna* brain coupled with this kind of science nostrum suggests that it is . . . and "always will be a mystery".

But other peer reviewed and published research has provided some new and very insightful answers. For example, when two people or one person and an audience participate in what we call verbal communication another interaction dimension comes into view—if we watch carefully. When people communicate they don't just hear and process sounds. They engage in what is called *motor mimicry*. For instance, when you are intently watching a high jumper, your tendency will be to lift your leg in empathy with the jumper. When you are shown pictures of people smiling, electronic sensors will record your penchant to smile, even without your awareness. When the carpenter working on my garage described his accident with a skill saw, which nearly took off his thumb, I noticed all three of us who were listening, grimaced in perfect sync with him.

Another researcher in Southern California has developed what he calls Affective Communications Test (ACT)[12] to measure the ability of one person to influence another without words (or prayers, mantras or spiritual connection etc.). It seems that individuals with superior test scores on the ACT can create a mood or mind set in those around them in two minutes without a word being spoken. How? They provide subtle, body cues and gestures. Here the researchers are not talking about the obvious bodily mimic technique but cues that are so subtle that even the researchers have not yet given them specific labels. One of the most famous studies to reveal the influence of super subtle non-verbal cues was done by William Condon. [13] He used a four-and-a-half-second strip of motion picture film. He watched it repeatedly for over a year. Finally he saw the micro-movements, which came to be known as *cultural micro-rhythms*. The communication in the film, in the final analysis, had a rhythmic physical dimension, not immediately perceived by the participants. They were "dancing" without knowing they were "dancing." Subsequent research has shown that people may begin a conversation with different physical rhythms but if the exchange is successful they quickly agree on a rhythmic micro-movement. So-called mind readers, spirit-connected gurus and the like, are often gifted with these kinds of clever sensitivity. They are expert (sometimes without their own awareness) at reading rhythms and micro-cues. They may choose to call it spirit, vibes, auras or some other 'New Age' cognomen and their brain *nodules* grow accordingly, making it more difficult to see other more parsimonious and rational points of view. But when Condon's film is slowed down sufficiently anyone who is not totally *nodulized,* or stuck a metaphysical belief system, can see the obvious cues.

Another illustration: Around the turn of the century a German school teacher reportedly taught his horse, named Hans, to do difficult arithmetic. "What is twenty divided by four?" Hans would tap his hoof five times. Whenever Hans and his owner put on a demonstration there were full houses of astonished audiences. Additions, subtraction, even fractions, were apart of Han's amazing repertoire. Then a psychologist named Oskar Pfunger demonstrated that the schoolteacher was unconsciously (apparently) cuing Hans with subtle movements of his head and eyes. Clever Hans was not so clever with arithmetic when his teacher was behind a screen but he was indeed very clever at reading the body language of humans. If a horse can pull it off, why are we so reticent to doubt the clever *mind readers* who sincerely (apparently) are able to read the cues and promote *otherworld* prognostications?

As mentioned earlier, there are rewards provided by James Randi and supported by the Society of Skeptics for the slightest proof of something *extra* sensory or *super* natural on the part of mind readers or spirit-connected gurus. The reward has yet to be collected. The fact is, there is only one *nature* and that *nature*, in and of itself, can astonish us and embrace us in awe, if we are willing to abandon our infatuation with coincidental correlations and *nodulized* beliefs. Without *nodule* addiction there is no paranormal or supernatural; there is only the normal and the natural. If you call yourself a Naturalist you will agree: if you are a Super-naturalist you may have no choice but to disagree. Yes, there are questions that may at this moment be called mysteries but given time and the insatiable curiosity of human beings, they too will be answered unambiguously.

Admittedly there is a lot we humans still do not understand. But most of what we call *supernatural, medium channeling, visions, intuition, or other world communication* can easily be explained via neuroscience. However, many addicted *nodulized* believers do not (perhaps cannot) hear the end of the story. A few years ago a popular female medium in a *trance* began talking in Hebrew. It was said that this was absolute proof of her connection to another world.

She must be a channel for some other entity since she said she had no knowledge of Hebrew. Upon further investigation and the application of Occam's razor (the law of parsimony) it was revealed that she had an *unremembered* Jewish nanny. This nanny spoke Hebrew to her during the first several years of her life! Most, even average, human beings have incredible capacities to receive and store information but not necessarily recall it in everyday situations.

Another problem arises when intelligent, desirous people begin to store and repeat these kinds of anecdotal (coincidental) stories. Typically, without conscious awareness, they use the information to support their own projections or correlations found in their own lives. They become addicted to believing their own integrated mythologies and their *nodules* continue to grow. They repeat the coincidence or part thereof until it becomes a dramatic and believable epic. Or they attempt to provide mysterious convoluted explanations that cannot possibly be pinned to reality or consensually validated.

A '*New Age*' declaration may be heard: "If we can get enough positive vibes united we can create a hundredth-monkey phenomenon!" (This turned out to be a dramatic exaggeration). Yes, some 'New Age' writers have implied that with enough positive vibes, a kind of neutrality can be achieved wherein all things and all physical laws are suspended just like the *Celestine Prophecy* predicted. Just try and prove that one, or the one about how many angels can dance on the head of a pin. Another I have actually heard is something like "heavenly spirits pulling up while devils are pulling downward—and because humans lack faith the devils are winning—apples are falling *'because of doubt'* not gravity."

ADDICTION AND THE "NEXT BIG THING"

I must admit I really do have doubts. I doubt that there is an immediate answer to the obvious world dilemma we all face. I doubt that any world leader, no matter how bright and charismatic, can create a lasting solution tomorrow. I have <u>no doubt</u> that fanciful fictions, whether they are wondrous adult fairy tales or archaic myths, will ultimately only provide more confusion. I have <u>no doubt</u> that convictions born of *uber-belief addictions* are more dangerous opponents of fact than outright falsehoods. Further, I do not doubt that we are changing. More importantly, I do <u>not doubt</u> that we human beings, by choosing to begin thought REHAB (Neuro-*nastic*s) have the possibility of changing our brain's long established *nodules*. Science, like Religion, promises everything but so far, the facts have not been delivered in a form understandable to Mr. & Mrs. Down-to-earth. I have lived long enough to recognize the cyclical nature of most things. So, if I am really honest, I don't expect the next great scientific breakthrough to be the immediate solution. However, I do see that while it is possible for us to modify our addictions, it takes *wanting* to expend concerted, **will**-full effort, and like REHAB, commit extended time to grow new *nodules!*

In this world, dominated by so much doom and gloom, I see that it is possible to communicate with each other about a worthwhile future. Rather than hoping for a pie in the sky bye and bye, or to blandly exchange abstract *uber-beliefs* about other-world membranes and dimensions, I see it is possible to live day by day *'in the here and now'* knowing that the cycle will always include the full compliment of life experiences: live well—love much—laugh often![14]

CHAPTER 6

"RELIGIOUS—SCIENCE" & SCIENCE—SCIENCE

The retreat from dualism

Recently my musician daughter has been making the rounds with her new CD, singing and playing in venues, including some churches. I enjoy her songs, so I sometimes go with her. In one of the *New Thought Science of Mind church* gatherings, a woman verbally engaged me. She happened to be a physician and maintained that there is a growing concern among "New Thought Science of Mind believers that somehow the new fMRI *brain scan* science is diminishing, that which is popularly called *spiritual experience*."

True, Newberg, Damasio[1] and others have empirically demonstrated that our two and half to three pound physical brain is responsible for the creation of the subjective experience of *consciousness* and *spirituality*—"all by itself (with a mere hundred trillion plus neuro connections)—without the dualistic supra-link to some mystical nether-land or cosmic intelligence. "Because of recent fMRI research," my new friend off-handedly suggested, "all of the Religious science, Noetic science and New Thought type centers must quietly merge with modern empirical science while the more traditional churches, if they are truly honest, must adjourn." She was serious about this. "Furthermore," she adamantly went on, "If Ernest Holmes or any of the other eminent sages were still alive, I'm sure they wouldn't be hanging back, quoting themselves and repeating all the old material week after week. They would be investigating these new neuro-sciences and brain-scanning technologies."

I could appreciate her point of view. Hey, at least MNBF *(my new best friend)* was questioning and not just *following the party line*. From what I've read, Ernest Holmes was always stretching forward. He took people where they were (mostly immersed in one form of traditional religion or another) and brought them forward into the new sciences as they were understood a half-century ago. (Ancient and primitive by comparison to today's science) Furthermore, it does seem logical as well as scientific that if *material neuro-molecules* are enough for the emergence of *mind* to develop, there is no need for the continuance of '*fringe science nostrums*'[2] The Law of Parsimony must apply. Einstein often said the one thing that is incomprehensible in this universe is that it is all—<u>comprehensible</u>. By hook or by crook, it seems that some of us 'modern folks' still want to create mysterious, unfathomable paradigms with which to confound others and ourselves—even with the elucidation of hard science so immediately available.

In the minds of most cosmologists, the Big Bang[3] demonstrates that Homo sapiens' advance from simple neuronal-synapses to consciousness has had at least three billion years to evolve and likely some ten billion before that, so it is perfectly reasonable that evolutionary biology rather than unknowable mysticism connects us to the unfolding cosmos. This whole brouhaha is reminiscent of the reaction to Galileo: "OMG what will happen to our spirituality if the sun not the earth is the center of our solar system?" Or later when Keats (the poet) censured Newton for deconstructing the rainbow by splitting the rainbow colors using a crystal prism. Allegedly he protested, "You have destroyed God's gift to mankind, the beauty of the rainbow." Evidently, some of those who still want to hold to a dualistic universe are repeating this same kind of grousing.

"Apparently," according to MNBF, "some very astute people think state-of-the-art brain scans and the like, are a spooky specters foretelling a loss that will gradually change the understanding of what some say *makes us most human*" (i.e. *'The subjective belief in a transcendent non-physical other-world and the possession of an immaterial something: soul/spirit/élan vital.*') "Many," she went on, "in my own twenty-first century *New Thought-Religious Science* movement are explicit in their objective of what they call *preserving what makes us most human, (a soul?)* in spite of the ever expanding brain-science worldview."

For some other *Religious Science* believers it appears that the solution to the growing apprehension is 'science' without the word 'religion' attached. At the same time, some *dyed in the wool New Age advocates* are carrying their members forward with the promise that their *non-peer-reviewed science* is on the brink of a new break-through! Still others insist on pulling quantum science out of its physics context [ch 5 End Note 4] to produce their *new scientific evidence*. "Apparently some of these groups" she said, "mistakenly believe that the discovery of *ultimate purpose* is lacking in the new neuro-science. It seems that these *believers* feel that they must seek and use whatever means possible to discover something beyond *mere mechanistic systems*."

Indeed, even I know that the allure of a *transcendental something-or-other* has been a major motivator since the beginning of written history. ('Belief' does have survival value!) And now we have published scientific evidence that the experience of the transcendental or ineffable is located within the neural labyrinth called he human *central nervous system*. So now what do these protesters do with their hopes for a bliss filled eternity? My new best friend referred to them as "arrogant hopes".

At first glance the move from mainstream peer-reviewed science into the ever-receptive grasp of marginal science looks exciting. Especially when we throw in a word like *transcendental*. "We love a mystery! And the distant promise of some kind of solution that will involve an eternal paradise of one sort or another!" was MNBF synopsis.

But when dualism rears its schizoid-opinioned-head by subtly elevating the *mental* over the *physical,* such pursuits begin to lose their appeal to bona fide scientists. Obviously, this enticement—direct experience of transcendence (or what she called the *PT Barnum effect*)—places *subjective* evidence over hard science; then consensually validated empirical data goes unheeded or at least unacknowledged. Now, with this thought in mind, we should gulp and think twice. This kind of subjective priority guarantees the age-old *dualistic believer's* conclusion: There must be *a spirit-like-essential-essence of me* <u>and</u> *an external-omniscient-god-like telos (end to end all) all tied up in a concept we call 'higher consciousness'* or some similar vernacular. (Most scientists and curious people have been there, done that—no T-shirt.)

But here, please allow the injection of a few obvious questions: Are our subjective feelings enough? Don't fairy tales, childhood mythology and wishful thinking garble our perceptions? If we are truly seekers of veracity, shouldn't we require credentialed, peer reviewed, scientific consensual validation of our intuitions if we are going to claim the word *science* as part of a title description? Don't extraordinary claims require extraordinary proofs? For most rational persons the answer is "yes" and today—if our preconceptions (*nodules*) don't get in the way—the really big questions can be answered by peer-reviewed hard science. "The incomprehensible is rapidly becoming comprehensible".

However, even a cursory overview of brain-scan research related to religion and spirituality brings us to—OMG—"*reductionism*". For some, the ability to understand phenomena completely in terms of the processes from which they are composed is the scariest thought ever! But basically, the realization that such research is inimical to the dualism required by most mainstream religionists is what the uproar is all about. The insistence that there is something '*out there*' which contacts/connects the '*in here*' creates the dualistic discord. And, if we look closely, we see that true '*objectivity*' undermines this kind of thinking.

Some may insist that mechanistic or observable neuro processes are not enough to answer the big questions. But once again MNBF had a repartee: "Unfortunately," she said, "as almost any *Reverend* will repeat: Human brains are mechanisms like radios or TV dish satellites to perceive a higher order and/or to facilitate research of the '*spiritual realm*. Look closely-this is dualistic thinking! They don't seem to realize that our brains are sending and receiving all the time." In another *Google* article I found that our ever-evolving human cortex has also been likened to a '*cognitive prosthesis*' for the soul that evolved to receive divine signals from somewhere *out, over, up, in—you insert whatever preposition you want—there*.

My MNBF kept repeating to me, "The problem is that there is no scientific or even logical account of *'how'* this dualistic brain perception of the higher order works or 'how' the soul receives alleged divine signals—or from precisely 'where'. Instead such questions are most often relegated to the subjectively miraculous or mythical mysterious, which again finds familiar company with the nebulous nostrums of fringe science and other secret compositions but without scientific proof." She seemed slightly condescending in some of her assessments. But, I think she had a point.

When you have eliminated the impossible via empirical techniques, whatever remains, (however improbable in our well rehearsed belief systems) must be, at the very least, a closer proximity to this world's reality. What am I calling impossible? i.e. the metaphysical nostrums—which can only be believed or disbelieved, rather than empirically validated, repeated and confirmed by consensual validation. Myths may comfort but they do not inform.

Sure, some New Agers may say we "make it all up" but when I look via a brain scan and see that within 'the noble brow' (prefrontal cortex) the mechanism that "makes it all up" is glowing vibrantly while "making it all up". It seems obvious to me that the devices "making it all up" are IN THERE not OUT THERE! Logically then, if we would continue to evolve and make the unfathomable—fathomable and the incomprehensible—comprehensive then it would seem logical that we must seek to know more about the glowing neuro-machinery rather than ancient-world philosophical tautologies. Yes, "young men saw visions and old men dreamed dreams" but they were obviously the products of their own brain mechanisms.

As MNBF suggested: "Our brains are not just radios receiving messages from 'somewhere out there'. Rather the human brain is the entire radio and TV station rolled into one. It has trillions of transmitters and it writes and produces the whole shootin' match! If we provide our *station* with delusional propaganda that is what will be broadcast. If metaphysical nostrums are provided as the repeated script then that invalid gobbledygook will be the broadcasted message." Gradually, I began to understand and partially agreed with her. I then emphasized the need for valid *'content loading'* and took another opportunity to talk about Neuro-nastics!

ALTERNATIVES TO CONSIDER

Today's neuro-research demonstrates without question that consciousness, intuition, and thoughts of ultimate meaning are in fact the ongoing processing of a 'googol' (*infinite*) number of neuro-connections within—what some have called, the '*mere*' human brain. Reduced to *mere* brain tissue, with its' one hundred billion neurons and more than a hundred trillion neural connections,[4] rather than a mechanistic view creating a "loss of that most human quality", an exact opposite conclusion may be reached. My NBF put it this way: "It appears that Neuro-scanning research leaves no room for a dualistic *invisible nether-dreamland* where our personal ego consciousness will live on for an eternity in the confines of what is often described as '*Christmas present bliss.*' Oh, woe is us! What shall we do?" I thought this was perhaps just too much derision.

"Come on now!" My skeptical friend continued, "Is the new mechanistic brain-science really a such a threat? Must we avoid mainstream science at the cost of returning to medieval alchemy? Is it really necessary to transcend the '*mere*' physical and its celebration of the natural world in order to satisfy some sort of existential crisis?" I suggested to her, that some make the assumption that the mind or the self, in order to be truly dignified, must be ontologically something other than just a phenomenal physical body.[5] For them, the answer to her 'existential crisis' question is indisputably YES and therefore Homo sapiens embrace dualism.

Robert Milton Ph.D.

HUMAN BEINGS ARE UNIQUE

Paradoxically, on some subjective level I "feel" ambivalent. There are moments when I too believe human beings are something more, something very special. For example, we all "feel" better and actually get better (if we're sick) when our environment is "pretty". What is that about? Personally, it seems to me that we human beings have or are "something more" than the "parts". CISTO. (*Can I Say That Out-loud?*) For example when we receive acupuncture for various aches and pains, we get better. Yet, according to Michael Specter (<u>New Yorker</u> Dec 12, 2011) research shows that acupuncture <u>is</u> a "placebo"—and while it is a worthwhile field of research—placebos resemble *nothing* or at the very most are like *faith healing.* Apparently they are effective with purely subjective symptoms such as pain or anxiety. Specter says there is no clear evidence that physical diseases such as cancer, arthrosclerosis or organic disorders of any sort can be cured or measurably improved by acupuncture (or placebos)—period! But why do placebos (acupuncture) appear to ameliorate pain?

Notwithstanding human cynicism attributed to scientists, this year, Harvard University created a bone fide 'institute' singularly dedicated to the study and understanding of why and how *placebos* are effective with human beings. This BMA. But I'm glad they are attempting to ask and find out why the "power of nothing" heals.

Predictably, no doubt about it, empirical neuro-science is connecting the inside to the outside and the higher to the lower because for authentic science, dualism is unacceptable. There is only the ONE substantial whole. So, on one hand, if *uber-beliefs* insist on something more than the evolved natural physical world then we do indeed need to abandon today's physically based neuro science. GMAB (*give me a break*).

However, on the other hand, if 'Science of Mind' *fans* can accept the recent findings that consciousness depends on brain materiality, then what was considered '*mere*' *matter* is not so '*mere*' any more. Once we begin to understand that the reduction of the higher to the lower and the part to the whole is not intended to eradicate but rather to explain and celebrate! At that moment, mechanistic accounts of human consciousness will no longer be threatening because these neuro-accounts produce a fusion of what only previously seemed like two different territories. You may remember that Einstein spent his life seeking a *unified field theory* that would tie together what appeared as irreconcilable data. Clearly, "*integration of all that is*" motivates both science and the human quest for genuine meaning—thus providing a solution to the existential crisis mentioned above. Furthermore—and this is the nut that has yet to be cracked by dualistic theory—*if certain types of experiences, definitions, meanings or purposes cannot bear examination by empirical science then maybe we should rethink our hankering for them and our attachment to them?*

EXAMPLES OF PSEUDO-SCIENCE

The problem with scientific fads is that while they are at the height of their popularity no one wants to find, let alone look for, possible bloopers in the research design. In fact, in both science and religion, after a new paradigm is proposed, publications and popular journalism tilt toward finding positive results. By the time flaws in design and ideas shift, the paradigm has often become so deeply entrenched in our popular cultural fabric that it has a life and following of its own—an *uber-belief*. So that even when new contradictory data comes into view, few bother to look at it, let alone consider moving beyond. For example, in the early 1930s, Duke University became the hallowed experimental hall for an exciting new idea known as ESP (*extra sensory perception*). J.B. Rhine devised an experiment using what were called "Zener" cards. It was a special deck featuring 25 cards printed with five different designs. So-called *intuitive subjects* were asked to guess the symbol before the card was turned. Most guessed about 25% correctly. But one undergraduate averaged much better than chance; in fact he pulled off spooky streaks of getting nine in a row correct. The odds were calculated to be about one in two million and he did it three times! The stunning results were prepared for publication and became the stuff of urban (and rural) legend. But between 1931 and 1933 the *intuitively gifted* student shuffled thru several thousand more cards but never again had results much above what would be expected by chance. Yet, even today we still hear the myth of ESP at Duke, touted as if it was a proven fact.

Interestingly, the more spooky or *far out* the paradigm the more exponentially popular the pseudo-scientific mythology becomes. In part, publications with dramatic results sell (while apologies for false-positives are printed on the last page in small print). It has been noted that even legitimate scientific publications are less likely to published null hypothesis experimental outcomes—they too want to publish favorable results. In fact, it is becoming increasingly clear that publication bias alone is producing major distortions in all fields. J. Lehrer in the December 13, 2010 <u>New Yorker</u> reported one of the more classic illustrations of *'reporting bias'* in the testing of acupuncture in various countries. Between 1966 and 1995 there were 47 published studies of the procedure in the Asian countries of China, Taiwan and Japan. <u>All of them positive</u>! Proving beyond a shadow of a doubt that this was *ancient miracle medicine*. During the same period there were 94 tests of acupuncture efficacy in the USA, Sweden, and the UK. Only a little over <u>half of these</u> studies found therapeutic benefits. This kind of wide discrepancy suggests that experimenters substantiate their preferred hypothesis while closing their eyes to negative results. Could it be that our *beliefs* are a form of blindness?

MORALITY: A GOAL OF NEURO-SCIENCE

It seems that today's neuroscience is searching for an innovative, more basic kind of understanding of *modern processes* without the *meta* this or the *para* that. These scientists see the human brain as a phenomenal causal system in and of itself and not at all in need of fringe science to bolster its astounding functions.

Furthermore, since the human brain, fashioned from the interactions of genes and environment, is the source of ALL our actions and beliefs, <u>free-will</u> and human <u>morality</u> become critical venues for immediate research. And yes, these same neuro-researchers are concerned about such questions; in fact a whole new playing field is emerging calling for the interfacing of sociologists, politicians, philosophers and scientists. For now they are calling this new field of study '*neuro-ethics*.'[6]

Such developments mean that an added basis for responsibility and answerability must evolve. The threat of '*gods above will punish*' is no longer efficacious in the control of human behavior. Recent efforts to bring together ethicists, humanists, scientists, evolutionary biologists and others concerned with social policy reflect the broad implications of current research on the human brain. According to My New Best Friend, "Along with the dawning realization that we *really aren't so very privileged over nature* comes the debate on how best to shape human personality and behavior. Now that we are in possession of new qualitative clout to facilitate the development of our *'perfectibility'*, will we be able to plainly label and make transparent the integrity of our judgments and definitions of morality? Lacking the threats of divine judgment, will we—can we—conduct ourselves ethically? Clearly, with or without the cleverly disguised machinations of our *trusted public servants,* we must hold each other and ourselves accountable. The shaping of good behavior demands no less! Still, an unambiguous understanding that human motives and behavior are a function of mechanical causes other than nebulous *free will*, may allow us to become more empathetic toward others, as well as more efficient in our justice system's *twelve peer* accountability rituals."[7]

    Recently our whole legal and court system is undergoing a kind of needful revolution related to neuroscience. Currently we rely on a very antiquated criminal justice system. New science suggests that both 'Judges' and 'Juries' have meager aptitudes for calibrating the *'truthfulness'* of a given state of affairs. Neuroscience can, even now, provide a great service to society at large by demonstrating that human brains in and of themselves are deterministic.[8] America's famous Fifth Amendment has already submitted to

DNA technology. Neuro-imaging could be next to be used as a kind of *forced testimony.* No palpable final answers yet, but at least the questions asked by today's 'Justice system researchers' are no longer sourced in Theology.

THE QUESTIONS KEEP COMING

Rather than the strongly reductive nature of scientific explanation being threatening, can we accept the new science as an avenue to a higher level of humanness? Indeed *consciousness* will not suddenly be eliminated by neuroscience. Consciousness can only be lost to scientific scrutiny if it involves something categorically non-physical with a Woo Woo overtone. <u>Subjective assumptions</u> are exactly the kinds of intuitions that science, and neuro-science in particular, can test. At its heart, science is meant to unify rather than make dualistic metaphysical pronouncements. Will we, can we, accept that?

Reductionism may seem menacing at first but when pursued to basic understanding—it is not. Rather than continuing to chase after alchemy can we seek and understand chemistry? Rather than calling correlation—causation, can we abandon astrology and numerology and seek a new appreciation for astronomy and higher mathematics? Progressive human history is the story of change. In the next few months our socio-cultural base will begin to morph dramatically—like it or not. Notwithstanding the kicking and screaming on the part of some *nodulized* traditionalists, the new Neuro Science and Genetics have surpassed all of our contemporary belief systems just as surely as Science did in Galileo's day. Inexorably, day-by-day during the past few months, our understanding of our world and our place in it has undergone a revolution. <u>If</u> or should I say <u>when</u>, we are able to open up to the new facts—that make Galileo's facts look like pre-school finger painting—our beliefs and spiritual concepts will inevitably also change. Dare we look at the new neuro-data?

Because we have turned our telescopes skyward and gazed in wonder at the billions of galaxies that make our own galaxy seem like a single grain of beach sand; all of us—radical and moderates alike—have cause to celebrate. What has already spectacularly changed, like it or not, is our understanding. We now see in verity what Einstein could only hint at—it really is a UNI-verse—not a MULTI-verse. Because we have glimpsed the conceptual nano-nooks and micro-crannies of quantum mechanics at work within the *'harbor of consciousness'* and have been humbled by the unity of the smallest of the small and the biggest of big, we now see that our concepts of infinity were too small because we now see clearly that we—it—they—are all the same stuff!

PART II

APPLICATIONS

Currently a scientific uprising is taking place across our land to free the human mind from extraneous and fictional ideology. We can all participate and benefit from this revolution because now science is finally being put into common vernacular and made available to all of us.
"It is required that we participate in the travail and elation of our age—whatever its form—lest we be judged not to have lived."

CHAPTER 7

NODULES OF MODERATION

Is Moderation another myth?

The United States of America lost the misunderstood *War on Terror* not because of superior or inferior military force but rather because Politically Correct ("PC") American *moderates* didn't believe we could—lose that is. As a result the United States of America has gradually morphed into what the friendly international coalition dubs a *paper tiger*. Even our President's *'bended knee apologies'* in Egypt fed into this perception—notwithstanding the Nobel Peace Prize.[1]

In the last couple of decades a significant portion of *Americans* have accepted the popular and Politically Correct concept of *tolerance* and *moderation* (T&M)—particularly when broaching the subject of religious beliefs. For this group *tolerance* and *moderation* represent a kind of easy-going, trendy, fashionable way of life. It is an idea thoroughly embedded in the fantasy words: Politically Correct (PC). And, as T&M attitudes epoxy themselves to PC modes of thinking, *Americanized brain nodules* begin to grow—exponentially since the 2001 attack.

Paradoxically, after the horrific events of 9/11, most PC—TM Americans took pride in being kind, considerate, and friendly to the purported *more benign representatives of Islam* residing in their neighborhoods. Yes, according to a recent *popular* poll (Vanity Fair) 59% of Americans are moderates. And some would say since "we are the majority, and we are also tolerant, go ahead and build a mosque on ground zero," so say some of the strident advocates of PC-*ness!*[2] "It is the polite, American way", so also say some media overtures, "be *embracing* as well." Yes! This is the traditional American *nodulized* way.[3] After all, *they* will respect our religious ideologies—won't they? (LOL. CISTOL?)

Tolerance and Moderation (T&M), they say, is the means by which diversity can become the sine quo non and "we can all live in peace"—together. Repeat it often enough and while it may not be reality; it has become nearly ubiquitous in the minds of some of our idealistic American public.

Even though, the apple-pie core of PC American citizenry likes to think of itself as benign, *tolerant & moderate* (TM) in their advertised world outlook, a closer examination reveals some unsavory dimensions. For the most part, *TMs* can be described as polite citizens who don't *rock the boat* or create discomfort for those with whom they cannot agree. This depiction fits mainstream Democrats and Republicans. Ordinarily, TMs don't protest in the streets, they don't write caustic diatribes to their congressmen; most of the time they don't even cast ballots, unless some neighborly, fervent flag-waver arouses them to get up and away from their addictive *dumb-down-box* to vote. They just want to live and let live peacefully. Of course, at the same time such action by TMs will encourage others to be silent as well. If the students of Tiananmen Square stand against armored tanks they may deserve to get run over. Right? Daniel Pearl was just asking to be beheaded, wasn't he? IDTS! (*I don't think so*) To say remonstrators are "just asking for it" when they use words, placards or cartoons to expose absurdities is to abandon a basic American cornerstone: Rational free speech! Yet, almost the entire US newspaper media refuses to use free speech. What! Yes, it's true, *The Washington Post, the Boston Globe*, *Chicago Tribune* and a host of others, recently refused to run one of Wiley Miller's cartoons. Why? Because the drawings could be interpreted as "poking fun at Islamic intolerance"![4]

Think about it! When one group uses words and symbols and the other group uses murderous violence, it is reprehensible to claim, "Both sides are in the wrong." The moderate position may be placate, placate, placate—but that doesn't necessarily mean it is the correct stance. When imperceptive fanatics use dark-age brutality against those who speak their mind, it is shameful to see the victims being blamed. Yet *TMs, by doing nothing, do exactly that!*

CONFIDENCE VS OVERCONFIDENCE

Today, *moderation* is the operative word meaning tolerance for everyone and everything to our polite politically correct (PC) evolved voters. Citizens of the United States, it is said, pride themselves on their forbearance and willingness to be the world's *melting pot*. Actually, for almost two hundred years, *melting* worked pretty well—a near authentic Democracy flourished. Then after WWII a spanking new kind of nationalism emerged in North America. The US suddenly became the most powerful military nation on earth and paraded compassionate, caring motives in every global nook and cranny—along with exploitation. Surprisingly to some, no matter how many good-deeds Yankee *boy scouts* performed in the far reaches of the world, ultimately it was too little or to no avail—primarily because "they" thought our arrogance and greed were far more impressive than our beneficence.

A perilous pedestal was erected that perched the US precariously atop. As might be expected, international collisions began to reveal fractures and cracks. Nations around the globe discovered that Korea, Viet Nam, Somalia, Iraq and Afghanistan succeeded in knocking and rocking *the mightiest nation on earth*. Fissures widened and the now wobbly pedestal began to crumble. Almost overnight the US gained a new title: the "supercilious, overblown, rapacious, paper tiger." Today, almost every developed nation of our increasingly fragile world takes pot shots at the American origami feline. Terrorists, suicide bombers, roadside bombs, tennis-shoe bombs, and briefcase nuclear bombs, have become the USA's daily portion.

IDEOLOGY AND MODERATION

In the illustrious US of A the *uber-belief nodule* of *Political Correctness*—tolerance and moderation—has developed in a relatively short period of time. The "walk softly but carry a big stick" has given way to a subjective *nodule* that repeatedly touts Americans as being courteous, kind, helpful and friendly as well as the fair-minded guardians of the world. But unfortunately, our former friends and allies of the heart-worn-world see us as spoiled children. Worldwide feedback indicates that *we are* not characterized as a kind, considerate, helpful Boy Scout doing good deeds for the unfortunate "poor and tired yearning to breath free". No, today, the world at large has come to be repulsed by us, for what they perceive as either unfeeling naiveté or imperialistic warmonger greed. Undeveloped countries resent what they interpret as condescending handouts. This perception is further magnified by a so-called altruistic endeavor to bring a *superior form of government* to the world, particularly the Muslim East. Americans, it is alleged, hold the belief that Democracy—chiefly *American* Democracy—is the ideal for every nation on earth. Yes, if you'll step back, this allegation does seem to smack of conceit! Somehow our politicians failed to see that where Democracy was introduced to Mid-east countries it often provided a quick *oil-slicked* route to a quasi-Theocracy rather than Democracy.

For the most part, Arab countries have already submerged generations of children into a barely breathable quagmire of religious laws (Sharia). *Uber-Beliefs*, not rational logic was and still is, the basis of Madrasah education.[5] War in Iraq and Afghanistan, by Yankee Christian infidels, only buttresses taught *beliefs* and grows bigger *nodules* in the brains of young Muslims. Now that the US has, in the eyes of the Mid-Eastern world, lost these wars, the *uber-believing* children of the East will undoubtedly be even more inflexible in their homegrown *nodulized* viewpoint.

Moderates must face the unfortunate fact that the various so-called tribe-tyrannies in the Arab world really do perform a function: They keep most of their own radical, *nodulized-believing* citizens in check. (Yes, you are correct, this is a non-PC generalization!) However, by now it should be obvious that if US politicians were magically able to plant Americanized Democracy in Muslim countries and <u>if</u> their citizens could/would vote for a particular party to take control, a majority would vote for a radical, religiously-based (Islamic) political group. Why? Their *uber-belief nodules* coupled with the fact that they have seen the radical religious organizations freely handing out millions of dollars—U.S. greenback currency. They have seen the duly *democratically* elected and well-armed religiously fanatical soldiers protecting the local citizenry. To Muslim citizens, these groups represent the palpable Theocratic reality of Allah in action. For the people of these Mid-East countries, Theocracy is factual—Democracy is fictional. Furthermore, what we may now call '*a fully developed Theocracy Nodule'* provides reason enough to reject the corrupting *MTV decadent Democracy* offered by the West.

It should be obvious by now that this state of affairs cannot be remedied by military intervention—unless mass genocide is the goal. Actually, come to think of it, I have heard some TV Christian '*Armageddon bound'* evangelicals calling for exactly that. The sooner the better! But again, should we silence such idiocy? Free speech—no matter how batty you personally might think it is—should be an inherent dimension of any genuine civilized culture. But pleeeeease at the same moment LOL! If we make loony speeches or if we draw persons as cartoons, so what?! Offending is probably not really rational and it's not my goal either but . . . *come on man!* Isn't it lunacy to worship a God who would create a *nether-land blast furnace* to incinerate those who do not believe in virgin births or dramatic ascensions into heaven via clouds or white horses? If you are offended, you should check out your reasons. If I shout to the rooftops the "world is flat" and you are offended, you have a bigger problem than most loony bin residents. If you are truly <u>secure</u> in your belief that the world is round (or virgin births, flying horses etc.), my saying it is flat should elicit pity or even condolences because you've gotta think (your *nodulized brain* would not allow otherwise) I'm off my rocker as well as my Zoloft. I do defend the right of Islam Imams or Christian Evangelists to preach or teach crazy ideas as long as those who have more rational world-views are allowed to laugh out loud. MNSHO. Yes, I think LOL is a good prescription to myth-oriented lunacy—it sure beats murder—In My Not So Humble Opinion!

'BELIEF' & MODERATION

Undoubtedly there are some who would call themselves *moderate* <u>Muslims,</u> just as there are *moderate* <u>Christians.</u>

Mostly these are what are usually called '*the more reasonable people*' who have decided to forgo identifying themselves with absurd and/or *inspired scripture* doctrine—particularly Muslims living in non-Muslim countries. But make no mistake, the future, which all Quran reading Muslims and Bible reading evangelical

Christians envision (*moderate* & radical), is one where non-believers have been converted—or else. A literal interpretation of either group's *Holy Scriptures* will not allow for the sharing of *paradise* or this world. Because of *ancient 'holy scriptural' injunctions,* Muslims as well as Christians are obligated to convert or vanquish non-believers including so-called *moderate*s.

Suffice it to say, being a polite moderate does not guarantee the avoidance of tragedy. In point of fact, if given access to a view of history, being a moderate—"*neither hot nor cold*"—probably facilitates catastrophe. When a bully invades the playground and starts harassing, those who stand idly by while a browbeater destroys the spirit of cooperation and freedom, pay a horrific price. Without a single exception, such bullying action results in tragedy but the real cause is seldom considered.

Here I pause to ask a thorny question: Is it possible for one to be *moderate believer?* That is, can one be *sincerely accepting* of another point of view while believing a contrary one?

News commentators use the phrases "peaceful Muslims" or "*moderate* Muslims" in contrast to "fascist Muslims or "radical Muslims". These contrasting descriptions may lead some to think there really are two ultimately different beliefs within the Muslim world <u>and</u> within the Christian world. For example, if one is a really devout Christian believer, is it possible to be *honestly accepting* of other religious ideologies?

When TV host Charlie Rose interviewed Rick Warren (the leader of one of the largest Christian churches in America) and asked about "Christian exclusivity", the evangelical leader suggested that he was "open". Then Charlie quoted a phrase attributed to Jesus: ". . . no man cometh to the Father but by me . . ." and offered that such statements seem to make Christianity exclusive. The preacher retorted: "I didn't say it, but I do believe it because Jesus said it. If you have problems with exclusivity, take it up with Jesus."

SEPARATE REALMS?

We have all heard the claim (*uber-belief*) that the sphere of religious belief and its domain of spiritual words should be treated as separate from *real* knowledge and/or the factual world. The word *faith* is most often injected as an assumed rational explanation. And when you think about it, moderates are, more often than not, willing to acquiesce to this kind of logic. This is how *nodules morph and grow.* However, that doesn't mean that such compartmentalization or '*unhinging*' from reality is really logical, intelligent or acceptable in any way. Somewhere I read: "When one person is '*unhinged*' from facts we call it psychosis but when several people become '*unhinged*', most of the time it is called religion."[6] When the president of Iran—whose opinions are commonplace in the Muslim world—advises the United Nations⁽⁷⁾ that Israel and the US government orchestrated the 9/11 attacks, it is way past time to LOL at him and anyone who would follow him. JMO

Then too, when it comes to the word *belief*, it is like the Tom Hanks movie—*Joe and the Volcano* where his mystifying "*brain cloud*" suddenly stops his *thinking process* cold in its tracks. The belief *nodule* dominates. My dog, Mr. Dugan, may be said to *believe* that the conveying of his evening meal and my arrival home is a fait accompli. It could even be said that he *believes* that I will *always* bring him a treat upon arriving home. His evident excitement and palpable tail wagging make it clear that he has made an association between my arrival and getting his canine chow. He may be said to be a *believer* in the emanate return of the *bone convoy*. This is not what most religious leaders want to mean by *having faith*. They usually mean that even without ever seeing or tasting a bone, they believe it will still appear someday. Even Mr. Dugan would give up on that one.

However, for most courteous citizens, '*believing*' under a '*have faith*' guise is unassailable. The 9/11 bombers were educated and relatively affluent, yet undoubtedly most of them really believed the fantastic proposition that if they blew up a building containing 3000 human beings they would be the proud recipients of seventy-two nymphs in paradise. Their belief *nodules* triumphed! Yet, when pinned down, most *moderate*s suggest that concessions must be made to such *faith-oriented* gibberish simply because it is defined as *faith-based religiosity*. Undoubtedly, repeated *thinking/ believing* of this kind has over the years, fashioned incredible *nodules* in the brains of *moderates*. Such citizens are unable to address the most pervasive cause of conflict today: *Belief* in concepts that are unfettered by evidence, logic, or any rational connection to cogent thought. By any valid definition, these kinds of absurdities should be defined as lunacy and deleted or at the very least derided as we might ridicule a self-proclaimed scientist who insists the world is flat or that lightening is God's way of having a temper tantrum. Acceptance of illogic coupled with the concept that if you are defined as *faith-oriented-believing-people* you should be free to actively promote absurdities (even suicide) is just plain preposterous; yet currently, this idea appears be the undoing of our Politically Correct T/M Western civilization.

To adopt illogical, unfounded beliefs about medicine or geography would soon lead to straitjackets and/ or isolation. To grant absolution to those who have completely personal definitions of words like *infidel, spirit, paradise, martyr, woman and Jihad* is also crazy and should be universally denounced and/or laughed at by every rational person on earth.

Surely the idea that the Muslim Prophet literally flew up to heaven on a winged horse should be seen as fantastic gobbledygook but no more implausible than what has been ascribed to the literal resurrection and ascension of the Christian's Jesus. Of course, m*oderates* would allow that both are matters of *faith* and that the *faith-full* should be free to *believe* such things without evidence. The almost universal idea that *belief* can be set apart and made real by something other than evidence has rendered all religionists (Christian & Muslim et al) unable to confront the most diabolical cause of conflict in the world today. The concessions made to religious fanaticism have prevented the United States and its powerful Media from being united in the war against *belief-based-nonsense, which is where the core battle needs to be fought!*

If a perfect human history can be said to reveal an imperfect human nature, it may be that the desire to *believe* without evidence reveals not only the very best quality of child-ness but also the very worst aspect of adult-ness. When we add nuclear fission to our current history we have the "perfect storm" to bring civilization's end. That reality is about as far from ludicrous as we can get!

THE PROBLEM WITH MODERATION

If we are to save civilization we must acknowledge that well-mannered *moderation* is the cauldron in which extremism and horror can roil and come to a boil because *moderate belief* provides the context in which *radical belief*, *faith-full extremists* and other absurdities **can and do** thrive. For example, most fundamental-religionists in America appear single-minded in their attempt to undermine the contemporary scientific community. Prudent moderates do no such thing. However these same sensible moderates make the world safe for radical fundamentalism by suggesting that "faith is an acceptable, even desirable, quality" and if undermining modern science is a part of that faith then good PC citizens of America must be tolerant of such belief-based activity.

The courteous concessions made to *faith-filled religionists* by moderates, mass media and polite politics have prevented criticizing, much less uprooting, the most inexhaustible source of violence in human history: Belief in absurdity. *Vis-à-vis:* The world is not flat! The earth is not the center of the universe! Devil spirits do not cause infections! Women are not property! Children must not be sold into marriage. Sending kids off to be suicide bombers must stop happening. Answers can be found in logic and research not by sending a witch doctor on some journey to a make-believe spirit world. Most thinking people deride, laugh at and desire to extirpate these beliefs—why not the rest of the nonsense?

Not installing rational focus—to expose the absurdity of inane ancient mythology—is an invitation to those who would assault and plunder civilization.[8] It is the proverbial public parking lot in a crime-ridden neighborhood where lights are prohibited. "But", say the modern moderate mediation experts, "positive thinking and love can do miracles in dealing with conflict and irrationality." I contend that *real* love, t*rue* love, agape, Philos or Eros would not, could not, exist for long in the calm, mild-mannered domain of moderation. When couples say that they "simply never have heated disagreements," I ask, "Can you make physical love without *friction?*" Friction, not immediate, automatic reciting of soft-spoken moderation, can, and probably must, be felt before love can be entirely fulfilled in renewal. Love of *self* and the *other* is central to any definition of a deep relationship and differences are an integral part of this recipe. As I said in chapter two, whenever you see cooperation be prepared to experience conflict. "Within every experience is the seed of its exact opposite."

SOME OTHER OPTIONS

In the 1950s, brainwashing carried on during the Korean conflict was a relatively new concept for the general public. At about that same time research into sensory deprivation[9] was begun at several universities in America. They found that when subjects were left in a closed container filled with a saline solution with no access to sensory input (obstructed ears, eyes, touch, smell, and taste) many subjects reported experiencing "an encounter with God." Apparently human beings have a genetically inherited capacity for entering a unitary mental state wherein we feel one with the universe. (e.g. deep meditation) Until recently this state was interpreted as the presence of an ineffable divine power. Because of modern brain scan research, we now know that when stimulation or even interaction with *another* is not available (picture a guru on the mountain top) visions of some supernatural 'god'-power may come about because apparently early-on we developed *uber-belief nodules* that comfort us by suggesting the source of our own thoughts come from an external power watching over us. So something that actually is initiated within us via firing neurons gets misidentified as coming from some divine outside.[10]

But even with strident declarations of alleged divine inspiration made on every street corner, it's time to call modern <u>*moderate*</u> PC religiosity exactly what it is: *Another sensory deprived mental aberration.*

To do otherwise is wussiness. IMNSHO This kind of sentimentality will destroy an authentic <u>civil</u> civilization!

Today's *T&Ms* and *extremists* are not at opposite poles. A closer examination of their so-called contrasting beliefs reveals a continuum. True they can be distinguished by the degree they invite political and/or military involvement. For example, the more political connection, the more uncompromising the *moderates* become. The more military involvement the more rabid the *extremists* become. On the other hand, can this be seen as an either-or? Like trying to be a "little bit pregnant," _moderate religions_ try making allowances for current-day literalizing of ancient scriptures. As a consequence, they actually continue giving birth to Terrorism. How? By creating a comfort zone for absurd beliefs to germinate, flower and propagate.

Moderate acceptance of these kinds of religious issues—like placating the playground bully—only facilitates chaos in our modern t*error-ridden* world.

Think about it! In what other field is literal, unquestionable adherence to outmoded medieval thinking tolerated? Medicine, physics, geography, astronomy, law and even politics have all moved forward. If a modern teenager, Muslim or Christian, were able to question the father of Christopher Columbus about any of the above disciplines, the senior Mr. Columbus would seem like a nincompoop. Today's kid would run circles around him in astronomy, medicine, physics and geography. But Columbus senior would know everything today's kids know about moderate fundamental Theology (i.e. the common-every-day-run-of-the-mill traditional stuff); his religious thoughts would be up-to-date and beyond reproof. Why? Because fundamental religion (Christian, Muslim or other) is the one area of human thought that does not allow for progress. The fables and old-wives-tales received by so-called r*evelation* or via modern *Sages* (of various names and spiritual qualifications) are considered as *holy* directives by *uber-believers* and remain untouchable by politically correct moderates.

It is imperative that we speak plainly about beliefs stemming from alleged *revealed holy writings* or *spirit-filled revelations* as silly or illogical. If we cling to moderate—*'let's-not-get-involved'*—attitudes, our children and our children's children will go on killing each other over so-called inspired *truth*. Because while a young radical Muslims may find bliss in suicide and fundamental radical Christians may rejoice in facilitating Armageddon, in a few short months even these kinds of choices may be unavailable to any of us.

CHAPTER 8

THE *NODULES* OF INSPIRATION

Epiphanies are accessible to receptive brains.

Human history is full to bursting with homilies demanding the acceptance of *inspired* books as not only *literal* but the *only* bases for salvation and/or patriotic fervor.[1] Unfortunately, the two—salvation and patriotism—have been entangled almost from the beginning of human history, which is why America's Founding Fathers made such a big deal of *separation* in their initial declarations and documents.[2] Today, current attitudes have once again allowed the two to become entwined. The matter of unscrambling the two—hopefully, once and for all—will undoubtedly wind up in the US Supreme Court—again.

Ancient writings that lend doctrine, dogma as well as what can be dubbed political policies, are considered by major religions (and many minor ones[3]) as absolutely central to their existence. Yet, we observe in the passing of history, that these same alleged "God-given, Divinely inspired compositions" have resulted in an assortment of belief systems within and without these same religions. The two major world faiths have over a billion followers and—implicitly or explicitly—claim to have the '*truth*', which is most of the time recommended as <u>the</u> way (the only way) to salvation, nirvana or paradise. Is this an arrogant POV? (*point of view*)

However, it appears that as the *information revolution* spreads around the world and popular science-oriented education is made available with a touch of a *Google* button, more and more citizens are seriously questioning *inspired writings* that insist:

- The world was created in a literal six days 6 thousand years ago.
- A boat, with every species on board, survived a worldwide flood.
- The female gender is inferior and need to be subject to male domination.
- Supernatural Deities require acts of sacrifice and specific rituals.

Instead, modern seekers of *truth* are more than a bit distrustful about *inspired writing* or *infallible utterances* made by historical and/or contemporary religious leaders—no matter how kitschy their clerical outfits. These modern seekers are more in keeping with Karen

Armstrong's[4] research suggesting that the ancient words and doctrines were man-made and inadequate. In fact, she says that the ancient clerics "devised spiritual exercises that deliberately subverted normal patterns of thought and speech . . ." OMG! (That *nodulizing* process was also elaborated in Chapter 1) And yes, that idea is certainly in keeping with modern fMRI research. There is a new breed of young educated inquirers coming on the scene who appreciate modern technology and yet they still want to be involved with like-minded groups *and* be of service to their communities. Sometimes when they discover that already established religious organizations could be an efficient way to accomplish these aspirations, they get involved in a local church community. For them, curiosity, commitment and service are the modern equivalent of "old time religion". They may even repeat or reenact age-old rituals, not because they believe a god introduced them to their church, but because they want to make a 'show' of involvement, and want their progeny to have a sense of the experience such involvement provides.

In previous generations, each of the major religions required total submission to what were called "*inspired written dogmas*" or at the very least deference to appointed interpreters. Today however, blind obedience and/or accepting on faith written revelations are not primary in the make-up of this curiosity prone *new breed*. In industrialized nations the mental repertoire of this "computer literate generation" contains a different view. They see the awesome feeling of new discovery standing in juxtaposition to fixed rituals and dogmas. In fact, many suggest, it is impossible to be fixed by explicit religious dogma and at the same time be open to new emerging input. (see chapter 2) Modern seekers of spirituality see it as encompassing child-like wonder, curiosity and new thoughts. These qualities produce the <u>awe</u>-*some-ness*, which is at the core of their *spirituality*. Religion, they say, is of a concrete and visible nature, while spirituality, like *'wonder'*, is abstract and like the wind, its effects are felt but more often than not, unseen. It is a fact that wind, breeze, zephyr as written in ancient manuscripts are words often translated as *spirit*.[5]

CHANGING ATTITUDES REGARDING DOGMA

Some months ago I attended the ordination of five Rabbis. What made this ceremony remarkable is that all five were women. Historically the Jewish religion has been considered as one of the most rigorous and restricted. Today there are a growing number of young moderns who know the Torah, observe *kosher* concepts and in certain settings communicate in the Hebrew language. They are meticulous participants in their faith but they also pick and choose. They openly pass over so-called *inspired scriptures* that fail to fit their educated modern sensibilities. Obviously conventions that restrict the position or function of women in their worship services are tossed out. They also openly reject scriptural teachings that imply that Judaism is the one true religion and that other faiths are inferior or lack the *'truth'*.

Still we hear the disapproving diatribe from the extreme orthodox personality decrying this modern change. *Orthos* (straight ones) see personal selection or change, in all of their multifaceted dimensions, as irreconcilable with commitment. *Nodulized*

Orthos seemingly cannot entertain the possibility of personal freedom being reconciled with the historic—therefore the singular orthodox—way to spiritual fulfillment. As a result of their life-long 'practice' they have *'straight'* brain *nodules* and are unable to bend. Their brains are fixed and they experience the synthesis of personal freedom and religious dogma as impossible opposites. They cannot help but rant about tradition, inspired sacred words, denominational roots and God's will. But, more often than not, they are paying lip service to what their parents taught them or a teacher's indoctrination and certainly, peer pressure plays a major role. Since they have been taught to believe before they have been taught to think, their rant becomes a growing ritual in itself.

However, when push comes to shove these same *Orthos* will often choose personal advancement over their publicly announced commitments. They quickly accept a step up the corporate ladder and uproot themselves from their communities. They will often discard conventional business regulations (time-honored ethics) they find binding or in conflict with "business is business." They may even divorce when marriage becomes unpleasant and thankless. They fall away from their '*ortho centers*' when they find the cleric lacking in entertainment. (Yes, I know . . . generalizations.)

But emancipation from fundamental *Ortho* religion and repetitive ritual is not easy without, at the very, least a neuro-NASTIC input or two. We have all seen the rage of religiosity morph into a tinkling bell or some other hollow sound. Historically, you can see it repeated in charitable organizations; that which begins as an altruistic *mission* eventually becomes an organized *business* with high-sounding mission statements, which inevitably evolves into a scam of one sort or another. (Yes, I know . . . more generalizations.)

At the same time, the myriad of new choices in modern life can be overwhelming. Like the guy who sits in front of his flat screen HD zapping the remote control. A whole evening may pass during which time dozens of entertaining programs are selected and rejected. There must be something else—better, more engaging, more fulfilling. Yes, just keep clicking; it's got to be there somewhere. The classic story of unlimited choice is making it rounds again on Email with one slight addition:

A store that sells new husbands has opened in London, where a woman may go to choose a husband. Among the instructions at the entrance is a description of how the store operates:

You may visit this store ONLY ONCE! There are 4 floors. The shopper may choose any item from any particular floor, or may choose to go up to the next floor, but you cannot go back down except to exit the building. So, a woman goes to the Husband Store to find a husband. On the first floor the sign on the door reads:

Floor 1—These men Have Jobs and Love Kids.

She is intrigued, but continues to the second floor, where the sign reads:

Floor 2—These men Have Jobs, Love Kids, Extremely Good Looking and help with Housework

'Wow,' she thinks, but feels compelled to keep going.
She goes to the 3rd floor and the sign reads:

Floor 3—These men Have Jobs, Love Kids, Drop-dead Good Looking, Help with housework, and have a Strong Romantic Streak.

'Oh, mercy me!' she exclaims, 'I can hardly stand it!'
Still, she goes to the 4th floor and the sign reads:

Floor 4—You are visitor 31,456,012 to this floor. There are no men on this floor. This floor exists solely as proof that most women are impossible to please. Thank you for shopping at the Husband Store.

PLEASE NOTE:

To avoid charges of gender bias, the store's owner opened a New Wives Store just across the street.

The first floor has wives that love sex.
The second floor has wives that love sex, beer and have money.
The third, fourth floors have never been visited.

Of course this exaggerated Email is not to be taken seriously since it is meant to be humorous. I'm pretty sure most men will LOL. However, almost every rational person would probably agree that men, not women, are more fickle. But then, this story/joke does dramatically reveal an almost universal dimension of the human condition: As a rule, we all want more—especially when it looks soooo easy to acquire.

INSIDERS AND INSPIRATION

Buddha, Jesus and Mohamed purportedly made many profound remarks. It seems obvious that most *inspired* prophets have attempted to pass on to followers *received* inspiration. How to accomplish this? It is well documented that the founders of our popular world religions were not the direct authors of the scriptural text moderns now possess and the founder's insightful utterances were *remembered* by others and then written down. Furthermore, most of the time more than a hundred years passed before the *sacred-writing* could be written, compiled and made available via the common written verbiage of the day. Yes, the words were written. But obviously most reasonable people would agree with Karen Armstrong, that human expression and/or holy symbols are totally inadequate to describe or contain what "*believers* conceive as supernatural inspiration". One way to accomplish inspired translation would be to receive tables of inscribed stone from the hand of Deity. Such words, it has been suggested, engraved by the hand of a Supernatural being would most likely be divinely inspired. However, established research reveals that the Mosaic story, was written at least a hundred years after the Sinai mountain performance. Yes, a <u>man</u> wrote the story of the "stone tablets given by Jehovah's hand" more than a hundred years <u>after</u> the event. *Inspired* though Moses may have been, the words of the Pentateuch in and of themselves were left to transcribers and translators. Moreover, the "thou shall not" rules that were reportedly passed from Yahweh to Moses were already a significant part of other Mediterranean societal codes. Such research suggests a bit of plagiarism going on shortly after the death of Moses. (Confucius uttered the golden rule five centuries before the New Testament was written. OMG . . . what does that suggest?)

It is most likely true that some men and women were considered '*inspired*' by their sycophant supporters but certainly their inspiration would have been put into words or symbols for the mind of those particular followers at that particular time to be able to grasp. Their specific words may or may not make sense to people of other cultures or other times.

From Sanskrit to the old English of Shakespeare's time, there are words and expressions in written historical manuscripts that are misleading because the meanings of many words change over the years. For example the proclamation: "that is the exception that proves the rule." For most, this well-known statement is supposed to mean that the exception strengthens the rule. But is that logical? Look again, how can an exception <u>prove</u> a rule? Obviously the word '*prove*' must have some other meaning. Yes! In an earlier time the word <u>test</u> and <u>prove</u> were synonymous. The US Marines still have *proving* grounds (or testing grounds) for exploring the consequences of their latest equipment. Now the statement begins to make sense:

"The exception *TESTS* the rule." Or as I mentioned in a previous publication,[5] the word <u>prevent</u> is currently thought to mean to *hinder* or even to *stop* but in 1611 it meant to "go-before" (i.e. Pre-vent). Having a heavenly street for Christians paved with gold makes sense only if gold is of value. Actually, early on in language evolution it could even mean something like yellow urine as in Goldwater. *'Meat'* in the Middle Ages meant any food (as in meat and drink). Today, it typically means only flesh.

CONSEQUENCE OF "INSPIRED" BELIEFS

Recently I received an E-mail regarding a mechanical engineering professor at Michigan State University named Indrek Wichman who sent an Email to the Muslim Student's Association in response to their protest of the humorous Danish cartoons that allegedly portrayed a caricature of the Prophet Muhammad. Professor Wichman's E-mail said the following:

Dear Muslim Association:

As a professor of Mechanical Engineering here at MSU I intend to protest your protest. I am offended, not by cartoons, but by more mundane things like beheadings of civilians, cowardly attacks on public buildings, suicide

> *murders, murders of Catholic priests (the latest in Turkey!), burnings of Christian churches, the continued persecution of Coptic Christians in Egypt, the imposition of Sharia law on non-Muslims, the rapes of Scandinavian girls and women (called "whores" in your culture), the murder of film directors in Holland, and the rioting and looting in Paris France.*
>
> *This is what offends . . . an academic like me and many, many, many of my colleagues. I counsel dissatisfied, aggressive, brutal, and uncivilized slave-trading Muslims to be very aware of this as you proceeded with your infantile "protests." If you do not like the values of the West—see the 1st Amendment—you are free to leave. I hope for God's sake that most of you choose that option.*
>
> *Please return to your ancestral homelands and build them up yourselves instead of troubling Americans.*
>
> Cordially,
I. S. Wichman,
Professor of Mechanical Engineering

Yep! As you might have expected, the Muslim group at this American university demanded Wichman be reprimanded for his attack on their interpretations of *holy inspired* writings. After all, the Allah "inspired"

Quran specially forbids such cartoons! (Or does it?) This Email story rather dramatically illustrates that America's war, is not with a country—not even with extremism, but rather with an alleged Divinely "inspired", *scriptural based* ideology that will not tolerate other points of view. Wichman's E-mail may have some politically incorrect overtones. IMNSHO. But isn't it time to admit that rival scriptural based religious writings (Christian as well as Muslim) are untainted by evidence of ties to today's reality? In fact, most fundamental Muslims (as well as some fundamental Christians) maintain separatist attitudes and generally insist on staying out side of mainstream society.

HOLY BOOKS AND MODERNISM

In America, religious *modernism* appears to be rapidly gaining ground as a kind of rebelliousness, whereas in Europe, *modernism* (liberal/intellectualism) is looked upon as a kind of relaxed, halcyon ideal. Most *moderns* insist on, (at a minimum) a noteworthy step away from *Holy Book* literalism. Certainly American *modern moderates* desire to avoid controversy and apparently most also desire metaphorical interpretations of certain *Holy Scriptures.* For example, while stoning people for heresy (as both the Bible and the Quran demand) is not particularly fashionable with moderns and most other religionists in America, such is not the case in other parts of our world. Google will show you several pictures and even videos taken in Muslim countries showing the actual stoning of human beings—mostly women. The penalty for blasphemy is death in the Old Testament. Nonsense you say? In today's Pakistan the penal code still prescribes <u>death</u> for *blasphemy* and the democratically elected government has enforced that law. Even more ridiculous: The 'crime' of blasphemy was against the law until 2005 in Britain. A few years ago a Christian group in London used the law to bring a charge of *blasphemy* against the British Broadcasting Company.[6]

Many contemporary middle-of-the-road Christian *'believers'* want their *inspired* religious writings to be more compatible with up to date science in general. Even though much of current scientific data contradicts the *books* upon which all—*modern* and fundamental—*Biblical beliefs* rest, most modern religionists believe they can have their *beliefs* and digest them too. They *treat* the Christian Bible the same way Thomas Bowdler did the writings of Shakespeare, by leaving out the so-called unsavory parts. (Actually, the *'bowdlerizing'* [*sanitizing*] was done by his sister Henrietta.) For example, most *moderns* would not equate illness with sin or demonic possession, as do the New and Old Testament scriptures. Selective scriptural perception is endemic to *modern* religious thinking and reading. Large sections of Christian scriptures are deliberately left unattended in most *progressive* churches. Like the eponym *'bowdlerizing'*, *moderation* may also be associated with censorship. This way, one may be called a *Bible-believer* and still sleep soundly at night.

I was invited to a farewell dinner a few months ago. Most of the other guests were *missionaries* of various religious organizations so dinner conversations involved mission field travels, anecdotal stories of miracles and such. One guest talked of an old *vertical church model* wherein various protestant denominations were separated by doctrines, biblical rituals or notions such as predestination, baptism, washing feet, anti-Catholic themes etc. He suggested that in recent years the ecumenical movement has created a new homogenized h*orizontal* denominational paradigm. "This," he said, "was accomplished by a more realistic view of Biblical interpretation. New *horizontal* Christians recognized that Biblical words couldn't be taken literally. For example," he said, "foot washing no longer has any relevance to modern life but underlying that service is the universal principle that does have relevance: *humility and service* to others is just as applicable today as it was 2000 years ago." He went on to suggest, "It is not the <u>words</u> that are *inspired* but the <u>men</u> who were inspired—to whom the various writings were attributed. Therefore the Bible must be read and interpreted in terms of <u>principles</u>, not in terms of the specific words." (Amen.)

Among thinking moderns, adherence to so-called *inspired* religious writing has evolved. C*hange* has not only been considered but incorporated. Because modern thinking people continued to question and doubt the accepted ancient "truths" and the *so-called factual data* of the that ancient time; *alchemy* became chemistry, *astrology* became astronomy, *numerology* became higher mathematics, *death by devil* became micro-biology.

REPEATING THE BASIC PROBLEM

If you please, I would like to repeat myself: We must acknowledge that well-mannered horizontal *churches* or even the *modern: "God is whatever you conceive churches"* can become the cauldron in which religious extremism can roil and come to a boil because <u>such activity and belief</u> provide the context in which *absurdity* survives.

Again, many *fundamental-religionists* seem single-minded in their attempt to undermine the modern scientific community. Sensible religious *moderns* do no such thing. However these same sensible *moderns* make the world safe for extremists and radical fundamentalism by suggesting that *'faith'* is an acceptable, even desirable, quality.

The courteous concessions made to *'inspired religion'* by moderates, mass media and polite politics have prevented criticizing, much less uprooting, the most inexhaustible source of violence in human history—<u>belief in absurdity</u>. The world is not flat! The earth is not the center of the universe! Infections are not caused by the devil!! Women are not property! Etcetera. The modern horizontal church mentioned above derides and extirpates these previously called *inspired holy writings* and beliefs—why not the rest of the mythical non-sense?

Not installing rational focus, to bring light to the absurdity of any literal acceptance (*moderate* or *radical*) of inane ancient mythology, is an invitation to those who would assault and plunder by using such acceptance as *passive permission* to continue their *'inspired' directives. (i.e. 911)* Yes, it really is like prohibiting lights in dark places. If it weren't so costly it really could be considered funny in some circles.

Notwithstanding, it really is time to call adherence to ancient so-called *'inspired writing'* exactly what it is—*Laughable*! Being hardwired to *believe* in order to *survive* is not the same as perpetuating magic as a survival technique in this day and age. When we are provided so many rational alternatives, continuing a *belief* in nonsense undoubtedly results in *nodules*, which result in thinking that is unassailable by logic. Obviously, this kind of thinking needs drastic REHAB (neuro-nastics) and empathetic 'therapists'.

Today's religious moderate *moderns* and fanatical *extremist*s are NOT at two different poles; their so-called contrasting *beliefs* reveal a continuum. (Can one be a little bit pregnant?) Accepting certain manuscripts as literally *inspired* and unchangeable continues to give birth to Terrorism by creating what has been called a <u>comfort zone</u> (albeit a horizontal one) for *absurd belief to take root and flower*. Acceptance or even toleration of literal *inspired* writing, is like trying to pacify a bully—it only facilitates more chaos in our already terrified world.

Please! Once again, think about it. In what other field is unquestionable adherence to outmoded medieval thinking tolerated? Medicine, physics, geography, astronomy, law and even politics have all moved forward. The fables and myths received by *literal inspiration* are considered up-to-date by conservative religionists of every stripe and color (Christian and Muslim) and therefore remain unassailable. Then too, *moderation* is a weak substitute for passion and enthusiasm, in a group setting, emotion will trump logic every time and believe me from what I have observed of radical religionists—they are passionate!

I agree with the implied conclusion of Professor Wichman's E-mail mentioned above: It is imperative that we speak plainly about so called *inspired beliefs* stemming from alleged *revealed* writings as laughable and silly.

CHAPTER 9

SURVIVING "AMERICANIZED" *NODULES*

Media sound bytes—bite back

There was a time when *surviving the coming crash* or *surviving inflation* were titles of best selling books. Today, simply *'surviving'* may be title enough. It seems the word *survival* is a high priority for the twenty five million recently arrived legal and illegal occupants to North American topography, as well as 275 million other officially entrenched but tetchy citizens.

For several decades fabled legends had the world believing that the golden road to *America* was strewn with the precious jewels of wealth and opportunity. Hollywood and Madison Avenue performed an extraordinary PR job in modifying almost the entire world's *nodules* vis-à-vis of the *American dream*. Especially dream—like was the *nodule* (growing by leaps and bounds as TV proliferated) that needed no electronically rigged device because it ricocheted through ubiquitous subliminal snippets: "*anyone can be President or an overnight millionaire.*" Belief in this kind of *Yankee exceptional-ism* has most likely morphed from the 1839 idea of *Manifest Destiny,* that is, the newly arrived white Europeans were said to be *divinely destined* to expand across the North American continent. That stunning bit of folklore killed off ten's of thousands of Native Americans then via *Manifest Destiny* took possession of their lands as if a WASP God had ordained it. This is where America's reputation for being imperialistic really started. TMI? It was a popular idea and still is among some, that America has been divinely sanctioned with a *special duty* to perform in the world. This seems a reasonable extension of the theory that America was founded under God and should have special duties to perform (e.g. *The only good Indian is a dead one*). Mike Kinsley of *Politico. com* suggests that such beliefs make it appear as a "boast that Americans are better than anyone else." Furthermore, such attitudes seem to be endorsed by a majority of American voters in that if a contender of the US presidency announces his exceptional "*Christian conversion*" he has a better chance of being elected. But belief in *exceptional-ism* has its downside because one of its major tenets is that "the '*other*' rules don't apply." Thus we invade Iraq and the UN becomes superfluous and when we lack the money to pay for *homeland goodies* and our attempts to reshape the world we Americans use credit cards. Why? Apparently we believe that our *specialness* is Divinely ordained. Self-sacrifice and the hard work necessary to make the American dream a reality are eclipsed by greed. Then the *Pollyanna* syndrome takes over and we vote for politicians who promise tax cuts, costly entitlement programs of all sort, plus the promise of a balanced budget and simultaneously keep building war toys. We are free to LOL when the next political promise is calorie free ultra-chocolate cake.

Such guarantees broaden the immigrant road and the world's poor and tired strain at the glistening gates of *America the Bountiful*. Then, most of the newly arrived visionaries realized, that like the yellow brick road to OZ, their ultimate destiny was entirely up to them. The *PR wizard* was all smoke and mirrors. They began discovering that the *nodulized* dream of the exceptionally '*good life*' for the mere taking was another *Grimm tale*. (Pun intended) Hollywood movies were exported all over the world and helped build the convenient-for-the-moment brainchild of a *free lunch*, which captured the ever-shortening attention span of the world's illiterate and destitute populations. Then TV continued to '*build it and they came*'—fifteen to twenty million strong without documents.

At a certain juncture in history just the word alone: '*America*' contained magic, myth and promise—these notions were and are abstractions that can and DO mean all things to all people. When emigrants arrive in tornado-like haste or from *Kansas*-like bleakness they have *nodulized* expectations—mostly unrealistic. Every citizen residing inside the *'imaginary borders'* of America needs to understand that separating fact from fiction is vital to their survival.

NODULIZING THE AMERICAN DREAM

Lately it seems that *survival* questions are quickly swept under or away by newer and more elaborate on/off analog questions. It seems no one is facing up to the reality of what has and is happening. All over the globe disasters constantly arise in which whole populations perish and others barely manage to endure. *Winners and losers*, they are labeled. But sometimes, quirky individuals will still ask one peculiar question and I think it needs an answer: Just who and what are the *winners*?

Within the buzz of our current *sound byte* most of the world sees the USA as a loser. They whisper things like, "The United States of America's time of supremacy is past." China and maybe India are currently perceived as the new economic dominators. The talking heads on Sunday night TV panels lend currency to the idea that all great empires must pass away. The United States has had its day just as the European Nations had theirs. The Twenty-first century belongs to Asia—they say.

True, you only have to glance at America's exported jobs record and the impressive economic growth of China or India to be reasonably certain that these are the *comers*. New, highly credentialed, but perhaps not completely unbiased, experts predict these rapidly emerging nations will triple their GNPs in the next ten or twelve years and a new species of turbo-charged capitalism of China will over-take the US as the world's leading economy.[1] Fresh 'takes' on old kinds of dramatic folklore regarding the goings on in China scare hell out of informed—and newly well-heeled—populations around the globe. Modern urban legends are being re-created on a daily basis by international news agencies. You may read, "every day a town the size of London springs up along the great inland rivers of China." And still another startling statement appeared in my web mail: "The West (and perhaps the whole world) has a new master."[2]

While there are some good reasons for you to be skeptical about world economic forecasts, it is not uncomplicated because there are so many difficult-to-deny facts being disseminated nightly for the public at large. Yet, I think caution is warranted in that economic statistics decanted from Asian autocrats have been notoriously unreliable. You probably remember the predictions made regarding Japanese world domination just a couple of decades ago, "What Japan could not take by force . . . they will buy instead." And they did buy Hawaii—almost.

A frightening fact that cannot be denied is that American economy is currently relying on huge infusions of cash from China and the Gulf states. But again, in the opinion of many, this may not signal the decline of American economic strength but rather it may portend the rise of new kind of proletariat—developing countries whose citizens have disposable income.

It does seem that nearly the entire world is benefiting from what is being described as a global boom! The average Joes of the world are beginning to enjoy a standard of living that was once reserved for the exceptionally wealthy and regal aristocrats. The myth of OZ is once again being disseminated—only now it is being applied to the citizens—Tin Men, Scarecrows and Cowardly Lions—of the Near and Far East. Common peoples of Communist China, the Gulf States, India and even Russia are participating in the Oz-esque mythology once reserved for newly arrived, naïve wannabe's at America's border. Voices in other parts of the world are now heard singing: "We get up at noon and start to work at one, take an hour for lunch, and then at two we're done . . . jolly good fun!" Globalization, which demands the *flattening* of the world, is part and parcel of the new mythological road to the *Emerald city* where everyone is relatively successful and living a parallel lifestyle.

There is no mystery about the "why" of US preeminence. Besides having enjoyed a *leveling* life style for nearly two hundred years there are at least three other observable themes running through the history of the *successful* United States of America:

1. The work ethic
2. Justice for all
3. Educational emphasis

These topics are not mythological. They are at the core of America's success—and any country choosing success.

MYTH BUSTING

On one hand it may seem like a fresh economic revelation has arrived from the new "boomer" nations namely: *poverty is crap!* Of course they want a better life—especially the life portrayed by exported Madison Avenue ads and Hollywood films. Their developing *nodules* demand no less. On the other hand, the basic reality regarding today's successful nation is really no different than ever. It centers itself in work, production and marketing. Work is obviously central to Chinese and Indian entrepreneurial triumph, just as work was at one time called the centerpiece of American heartland success. As every honest immigrant to the US discovers, there is no magic or mystery involved in realizing the "dream". Being constantly frustrated is often a miserable life style but as most find out, the real frustration begins when they get what they came for. Being documented and in the "system" carries a high price. Ninety nine times out of a hundred, the bottom line to achieving the American dream is a matter of hard work, long hours and taxes. Thomas Edison once said: "The reason so many American miss opportunities is because it is often dressed in overalls and looks like hard work." Luck can be a factor, but as my dad used to say, "The harder I work the luckier I get!"

WORK ETHIC

It has been said that the implementation of the "work ethic"—by rugged individualists—is what initially gave rise to the *American* dream. It is what made the country great—*they* say. In the past, *work* was not just an abstract paradigm but rather it was an idea coupled with rolled up sleeves, an honest day and an honest dollar. But it seems that recently the historically described work ethic has had a near fatal stroke—(too much cholesterol or lack of exercise?) For today's generation it seems work is waving one's hands while chatting or texting and/or a successful IPod download. What is needed today is not a *work ethic* but an *ethic* period! Hello . . .

It was *ethic* coupled with work that made America great—not just work. What *ethic* means is that your labor is useful to you <u>and</u> to other human beings. It means that there is a concern for others, a feeling for others as well as yourself, so what is called a *fair share of work* might not even enter into the equation. This kind of *ethic* says your share is not just a percent but rather it means it means you give a hundred percent. A marriage, for example, based on a 50/50 or even a so-called fair share arrangement is doomed from the start. Anyone, who really knows, will tell you that marriage, most often, requires a hundred percent from both partners. Lovemaking is a good corollary-example. If you contribute less than one hundred percent you will receive diminishing returns. In a fulfilling relationship one plus one adds up to a great deal more than two.

Furthermore, this kind of *ethical* effort usually has a quality of innovation, which can be very threatening to some already established work procedures. How? If your work has a creative bent you will, more likely than not, meet resistance. Why? It seems that the current large corporation workplace is bumping along in a state of uninterrupted torpor. The American auto industry recently collapsed in a state of languor because of the lack of innovation.

No doubt, in the process of *work/survival*, mistakes are underrated—making a mistake, admitting it and moving on is the real beginning of consequential *work ethic*. Mistakes and improvements are two more aspects of a meaningful work *ethic*: *Mistakes* (even if denied or undervalued) are a major survival objective and *improvement* is what ultimately matters—no matter how modest.

To be currently and continually relevant, the work *ethic* means that you imagine like an artist and create like a scientist. If your end work product is inventive as well as useful, that is the *scientist* at work. If the process is unpredictable and explorative, that is the artist. When insatiable curiosity encounters consistent attention to detail *a new phenomenon* is made possible—indeed, new *nodules* grow and new epiphanies are manifested. That is the work *ethic* in action.

Make no mistake; most of China's citizens work like there is no tomorrow. (Maybe they are right—*morrows* in China, by some reports, are tenuous to say the least.) Their reputation for work appears unmatched in the world but their lack of ethics is also unmatched. Their willingness to pollute their own country and the world is malevolent and will eventually prove to be suicidal. The lack of *ethics* in China is the primary reason many predict they will not reach America's eminence.

JUSTICE FOR ALL

The eighteenth century "*renaissance man*" and philosopher Rousseau[3] believed that government, hence law and justice arose through a *social contract*. He argued that humans have a natural disposition to empathize and eventually come to value the opinions of others. This state of affairs eventually led to the suggestion that the first State Government was basically a contract invented by society for the benefit of all and was arrived at by the people as a whole, who then voluntarily agreed to subordinate their own individual rights to the *State*. It was believed that only by allowing a neutral *State* power to dispense justice could "equal justice for all" be assured. Another POV was recently offered by Jared Diamond a modern *renaissance man*. [4] He equivocated the concept of the *social contract* by declaring that people do not typically voluntarily abandon personal rights and organize themselves into "State Governments" without external duress.

His research among the peoples of the New Guinea Highlands led him to conclude that what many people call *justice* is more closely allied with vengeance. That is, the human brain has certain evolved dispositions; the desire for vengeance is one of them. Diamond suggests that vengeance is a powerful human emotion ranking right up there with grief and fear. While modern State Governments make significant allowances for the expression of grief, fear and even anger, they mistakenly equate the expression vengeance with uncivilized, crude behavior. Furthermore, if a particular populace is to become civil, then this vendetta-oriented emotion must be left to the *State*. But, an individual giving over the right to an external authority for the exacting of personal vengeance is not particularly easy.

This is where Diamond observes that either (#*one*) external pressure from an encroaching power provides the impetus to give over personal rights to a *State* or (#*two*) continual warring between groups finally results in one group gaining ascendancy over the other and thus the victor becomes responsible for meting out justice—that is, providing an act of vengeance on behalf of individual clan members. Furthermore, if the *State* is unable to dispense realistic justice or muster credible demonstrations of force, personal displays of vengeance readily resurface—personalized vengeance then, may be more genetically hardwired than *nodulized*.

Giving up this probable *hardwired* personal right of vengeance, from Diamond's point of view, is neither natural nor particularly helpful to the human psyche. In order to persuade (*nodulize*) us to do so, State Governments and their embedded religions insist that vengeance and vendettas represent anti-civil, low caste types of behavior. At the same time, by allowing certain nameless *State* persona to exact the required revenge, we claim this is civilized justice. Further, this kind of State sponsored 'settling of scores' is *believed* to embody a higher form of evolved civility and when this behavior is repeated often enough and long enough, it will indeed become a learned *nodule* that is once again, like spirituality, taken for granted as God-given.

It seems obvious, to most people in our Western culture, that allowing individuals the personal right to exact vengeance (justice) would make it impossible to maintain a peaceful coexistence within the same community. Diamond makes the point that without the *State* in control we would begin living under the constant threat of clan warfare, which Diamond says is widespread in non-state societies like those of the New Guinea Highlands. In this same vein another Cambridge educated historian,[5] spoke of a deepening problem within Islam. It is, he said, a *world* in which human life doesn't seem to have the same value as in the West. In agreement with Diamond, Benny Morris declares that most of the Muslims are immersed in archaic *tribal culture* in which *revenge*—on a very personal level—plays a central role. A tribal society, he says, lacks what most of the rest of the world calls "civilized moral inhibitions". So that when they obtain chemical, biological or atomic weapons they will not hesitate to use them.

No doubt, most Americans feel that *State* laws, which preclude individuals from acting out revengeful feelings, are needful and should be supported. At the same time, perhaps making a clean breast of such feelings, not only needs to be permitted but encouraged. If, as civilized people, we believe in the lofty ideal of equal justice for all, it would seem logical that we also be willing to acknowledge the underlying feelings of righteous indignation and vengeance in order to insure the future survival of that ideal.

However this whole line of reasoning assumes that "the desire for vengeance is a powerful human emotion ranking right up there with grief and fear." But is it? One of the most prominent cases of modern-day revenge can be found in the case of Bernie Goetz—the NY subway vigilante—who when asked for five dollars by four rowdy teenage thugs pulled out a .38 pistol and shot at them. The consequence was a national sensation. All four had criminal records and even though one of them was paralyzed for life, America's public lent its sympathy to the mild mannered, *avenging angel* of NY urbanites, so much so that when he was acquitted, a spontaneous street party erupted outside his apartment.

An in-depth psychological study[6] of Goetz however, reveals a deeply disturbed man who was set for a violent explosion no-matter-what. So the question remains: Was revenge the motivation for the shooting? Diamond would have us believe that revenge is a normal human reaction and Goetz was only doing what is naturally hard-wired or perhaps *nodulized* in all of us. Of course, the more traditionally liberal explanation is found in the context of early psychological trauma or inadequate role models as causative factors for rage and vengeance. Actually a brain scan could tell another story of pathological brain circuitry. A virtual plethora of *bleeding heart* and/or scientific explanations could expose us to many possible reasons for vengeance. Bottom line: Bernie was obviously out for vengeance and according to Diamond, his response could/should be considered quite normal.

OTHER EXPLANATIONS

The more recent highly publicized *genomic* explanation—"we are all victims of gene collusion" represents a very popular justification. Genes, we are told, predispose us to violence, crime, revenge and possibly other moral failures.[7] Again, *free will* gets put on the back burner.

On one hand, there is the cultural *blame game*; the downward spiral of American society is said to be a function of social injustice and racism. In the same breath we hear that we need more Federal and State money to restructure economic inequities in the educational system and the employment market. Yes, we often hear that it is so obvious that *crime* and its cohort *vengeance* are the consequence of fundamental failure on the part of government. "They" therefore, it is said, "must undertake gargantuan and costly steps to remedy the situation." So *they* say. (Does correlation equal causation?)

On the other hand, it has been demonstrated that crime in NY and even the face-off between white Bernie and four black teenagers may have had very little to do with twisted psychological pathology, racism, poverty or the lack of parenting. The dramatic drop in crime rate in NY—and therefore diminishment of the desire for vengeance—was shown to be the result of recognizing and applying a remedy to the *Broken Window Syndrome*. When Mayor Giuliani and other

NY leaders began to fix the broken windows, wipe off graffiti, put an end to panhandling and peeing on the street—minor, seemingly insignificant everyday offenses—crime, *major* and *minor* diminished. (Pardon me, but does correlation equal causation?)

One conclusion from the NY results might be—rather than having revenge sourced in societal aberration, fixated historical tradition ("my grandfather and all my ancestors did it this way"), innate neuro-genetic motives, or even being hostage to a *nodulized* dysfunctional clan—that the average person is intensely sensitive to external cues. These cues and *perhaps* unconscious motives born of previous environmental perceptions are in and of themselves capable motivating human beings. Validation of these ideas can be seen in both *mob action* and the *broken window* syndrome. This argument suggests that criminal actions are the stepchildren of an environmental social context. In this context it would appear that *Neuro-nastics* could be the real answer to societies woes. So that while the NY crime stats may have had little to do with the traditionally accepted psychological rationalizations, they had everything to do with graffiti and minor public offences. [8] This explanation suggests that we don't necessarily need State Governments to flex their martial might or empty their tax coffers to end crime and vengeance or to ensure a sense of social justice. Rather we only need to scrub some walls and put an end to various public nuisances petite step by petite step. When order replaces chaos a sense of fairness prevails. Fairness precludes vengeance. New perceptions (*nodules*) prevail.

But this latter argument implies that we hominids are rational to the core and completely ignores the fact that NY "State Government" erased the graffiti, arrested and jailed the "minor-rule-offenders"—petite step by petite step. At the same time, it must be acknowledged that the "petite step" policy simply says that we are powerfully shaped by our neuro-perceptions of our environment—the color of a wall, the lighting on a concourse, people strolling or rushing—our physical world plays a huge role in how we behave. It does seem true that we behave differently (better?) when the street is clean rather than layered with graffiti or enveloped in garbage. Selective schools where students have simple dress codes have fewer behavioral problems. [9]In this context we are changing the cues that invite crime and malicious behavior in the first place. It seems important then to analyze and understand vindictive behavior. Perhaps, it is even more important to construct an environment that negates crime by glistening with fairness because this kind of creation can change a mental set and be shown to prevent vengeance and vigilantism. The implications of this uncomplicated idea are enormous. Researchers have collected a body of evidence[10]revealing, among other things, that perhaps kids are better off in a high-quality neighborhood with dysfunctional relatives than in a dysfunctional neighborhood with high-quality relatives. This POV BMA!

BACK TO THE FUTURE OF SURVIVAL

In our modern world, dominated by global markets on one hand and isolation and imperviousness on the other, if survival is possible, it seems vital to at least begin to separate fact from fiction—myth from metaphor—*nodulized beliefs* from actuality. In the past, simple survival was exactly that—*simple.* We depended on what was called *survival instincts.* Today these so-called instincts (common senses?) have been complicated by millions perhaps billions of other variables. But survival, whether or not we know it on an objective level, is a property of all kinds of cues, including human behavior, societal institutions as well as genetics.

After examining a couple of the ostensible raison d'être for America's success: The *work ethic* and *equal justice for all*, (Education will be addressed in another chapter) I would like to introduce some frightening arguments for America's ruin. American and therefore your survival, depends on you and your shared synergistic sense of an optimist outcome. Basic survival sense seems to insist on the anticipation of a *future*. When you think ruin and desolation, you help produce it. The *nodules* grow; you and your peers experience spoil and devastation. Think copious amounts of survival! That takes what is called WILL-fullness REHAB and much time spent doing survival mode exercises, and practicing survival type *Neuro-nastics*, your brain *nodules* change reflecting that same mode and the world will survive one person at a time.

CHAPTER 10

EDUCATING BRAIN *NODULES*

We know what does NOT work

When I was still in grammar school, my mother—who did not finish the tenth grade—would often say to me, "Get an education Bobby and you will never have to dig ditches." No doubt somewhere along the line she had done work that equated to digging ditches. She was a very energetic and powerful woman—so maybe she actually dug ditches. TMI? At any rate it was impressed upon me that education was a *must*. Sure enough I proceeded to take advantage of all the avenues of education available. From college to University graduate school, one door after another opened and I threw myself in with gusto. I admit that at that time in American history *Education* was indeed emphasized as the *sine quo non* by everyone. (Not just my mother) My generation, at least in the state and city where I grew up, seemed united in the dreams of educational accomplishment. Education was a prize carrot not a painful pitchfork. Today, especially since 9/11, many kids seem driven by pitchforks—common enemies, not common objectives.

'Founding Father' Thomas Jefferson's creative thoughts regarding the *Education process* need to be re-examined. (And more importantly perhaps we need to reexamine the education *nodules* that have been carefully tended and organically grown over the past two hundred years.) More than any other *Founding Father* he was willing to share his *nodulized* fantasy of what *Education* should be.

When JF Kennedy was president he gave a banquet for every living American Nobel Prize winner. (Approximately 150 were in attendance). JFK made a toast during which he said, "Never has so much genius been assemble in one room, since Thomas Jefferson dined here . . . alone." True, he was a brilliant statesman, linguist, architect, musician, agronomist, scholar, humanist and an articulate master of political stratagem. Mostly however, we remember him for his: "Government of the people and for the people," rather than his caustic evaluation of religion particularly Christianity. He said: "I have examined . . . the superstitions of the world and do not find in our particular superstitions of Christianity one redeeming feature. They are all a like, founded on fables and mythology. The effect? To make half of the world fools and the other half hypocrites, to support roguery and error all of the earth."

Of course, his reputation for owning slaves and fornicating was apparently also a part of his reputation. DISTO. (*Did I Say That Out loud?*) The chronicle of US Presidents having illustrious sex lives is well documented. DITTO *(yes, I did say it out loud)* Historians have long known that Thomas Jefferson, the third President (1801-1809), fathered a child by one of his slaves. Today we might ask, "was Tom, one of America's *Founding Fathers*, intentionally staging live sex shows at his stately Monticello home?" We know that Jefferson ordered the construction of a glass greenhouse and wooden verandas outside the master bedroom's south windows and he also built a glass-paneled structure near his study—featuring a cozy alcove bed. Was his idea to conceal or expose *her* presence in his bedroom? Incidentally, the house additions mentioned above were made the same year his alleged African mistress gave birth to their first daughter.

Notwithstanding, Tom's ultimate virility was more likely assured by his authorship of the Declaration of Independence and the Statute for Religious Freedom. But similar to his sexual feats these complex exercises in statesmanship were by no means accomplished without some embarrassment. History acknowledges that Jefferson did have collaborators but even when he embraced the epiphanies common to an author he was never completely satisfied with his final "climatic" *effort*. It just wasn't *'good for him'*.

His ultimate orgasmic ambition and likely his most exotic fantasy was an elaborate plan for making education available to every citizen, <u>and</u> a University education for gifted youths unable to afford the costs. He considered his most important accomplishment to have been the founder of the U of *Virgin*ia. Clearly he knew that no republic could maintain without educated civilians, which would *"enable every man to judge for himself, what will assure or jeopardize his personal freedom."*

His plan to pull off his educational fantasy included dividing every county into parcels, *"of such size that all the children of each will be within reach of a central school in it. Rendering the people safe—as they are the ultimate guardians of their own liberty."* Remember, the total population in 1775 was two and a half million people, and only about a third of those were apt to be urged into schoolhouses. (Probably much less, since women were still considered chattel and not allowed educational *'rights'*.)

He also made these rather remarkable statements a propos public and private education:

> "Education is here placed among the articles of public care, not that it would be proposed to take its ordinary branches out of the hands of private enterprise, which

manages so much better all the concerns to which it is equal; but a public institution can alone supply those sciences which, though rarely called for, are yet necessary to complete the circle, all the parts of which contribute to the improvement of the country, and some of them to its preservation."—Jefferson: 6th Annual Message, 1806. ME 3:423 "The object [of my education bill was] to bring into action that mass of talents which lies buried in poverty in every country for want of the means of development, and thus give activity to a mass of mind which in proportion to our population shall be the double or treble of what it is in most countries."—Jefferson to M. Correa de Serra, 1817. ME 15:156 "The less wealthy people, . . . by the bill for a general education, would be qualified to understand their rights, to maintain them, and to exercise with intelligence their parts in self-government; and all this would be effected without the violation of a single natural right of any one individual citizen." Jefferson: Autobiography, 1821. ME 1:73

A relatively free education then, was to be assured to every citizen <u>by the government</u>. We should also remember that models of education in Jefferson's day were committed to the ideal of laissez faire, defined as minimal government interference in civilian activities and/or wealth acquisition of citizens (numbering three and a half million).

While today we may consider the idea of public education a *slam-dunk*, it was certainly not the case when Jefferson was first advocating it. Like suffrage for women it took over forty years in America to bring his dream to a quasi-reality. For education to be accepted as a free entitlement, worldwide, took even longer.[1]

If we follow conservative ideology from the turn of the Twentieth Century through the Great Depression and New Deal and focus on the rebirth of conservatism under Ronald Reagan, we find *lobbying* as the major power in transforming our Washington capital into a massive orgiastic frenzy for corporate special interests. How did this happen? Answer: Greed and an uninformed, uneducated and indifferent public. Almost every incumbent government since (Republican and Democrat) has repeatedly run America into huge spending deficits in order to '*defund the other side*.' The end result: A government that appears concerned for its people but is in reality ineffectual and incompetent. (Proof that big government as well as big education is inherently inefficient and incapable no matter which side is in power.)

The United States of America—with its government "of the people and by the people" can survive only if the populace is able to be informed and stay informed. "If a nation expects to be ignorant (*uneducated*) and free, in a state of civilization, it expects what never was and never will be." (Thomas Jefferson in letter to C. Yancey 1816)."

What difference would it make if we substituted the word *motivated* for *informed*? Actually, for the motivated citizen today, there is a wide array of educational opportunity at the informal level—such as museums, libraries, social clubs, churches and most obviously the internet—this fact would seem to leave the *right* pretty much to the *motivated* individual to pursue rather than a top heavy bureaucratic government to provide. (MHO too)

Today, in many instances, day-by-day experience or what we call *street smarts* has become a matter of actual survival and for good reason has supplanted the basic three Rs as more essential for many urban citizens. As a consequence *we* have spawned a whole generation of near illiterate, but non-the-less school graduates. A famous American poet[2] once said, perhaps with tongue in cheek, (TIC) "An Education provides the ability to listen to almost anything without losing your temper or your self confidence." While this has the ring of a profundity, it could easily be a definition of a con man or a sociopathic personality. And yes, you're right, it does seem that more often than not, con men and sociopaths are the product of our contemporary urban educational system. Allow me to inject a NSHO (not so humble opinion): As *Jeffersonian education* morphed into an entitlement, a mental set or *nodule* developed which incarcerated *out of the box* thinking about education for a dozen generations. No other models of education were 'given the *time of day*'. And finally, when free education was described as a right by the International community in the 50s, the USA's old educational model was pretty entrenched in and around most literate nations.

No doubt, today most of these same nations see that they too must re-focus and bring new ideas to the educational table. New models are desperately needed but it seems that few are being actively sought today. Yet, for example, how can we get our thoroughly fixed minds round an idea like "education is a *privilege, not a right*"? When we have been immersed in the 'entitlement—*right' thought process* for more than a hundred years? A *nodule* indeed!

PLANS, POSSIBILITIES AND PROBABILITIES

Today's almost constant hue and cry: "Let's do something about our educational system . . . it is obviously failing", seems to have become another *cry-wolf* memorandum. The plea "to do something" is on the lips of parents, politicians, proletarians, bohemians and bourgeois alike and in spite of all the rhetoric, complaining, repetition and financial drain—nothing has helped so far. Why? Because Plan A has not been seriously considered, let alone implemented. "Plan A:

Our educational system must have a complete overhaul from the ground up!" There is no plan B!

And let's not kid ourselves; such an undertaking, if done by the usual political methodologies, would absorb all of our National energies and our entire National budget. It would make the recent *bailout* seem like a trip to a thrift store! True, we need a whole new system of education to meet the changing needs of modern citizens. True, the present system is a failure from the top to the bottom. Almost from the beginning of this current century our prestigious educational system has been driven by what could be called the Great Educational Ponzi Scheme.[3] A Ponzi scheme, as we recently re-discovered, is a '*con*'. It is a fraudulent investment operation that pays returns to investors from money paid by subsequent investors rather than from profit. It is based on the false premise that like '*hope*' *it* will never *run out*. Our educational system has been pushing students through the grade hurdles without consideration of what comes after they have the diploma in hand. Teaching *to pass a test* has absolutely no relevance to life coping or survival skills and the '*con*' has not only caught up—it has *run out*.

This unbridled exploitation of educational principles should not cast aspersions on the teachers who were just doing what higher powers mandated. No single entity designed our failing educational system, it just gradually devolved from population growth, corporate and political demands. As long as the population continued to grow, the 'Ponzi-Ed' system flourished. The next flood of Boomers or immigrants would take care of the *con*. For a long while, as systems go, it actually worked pretty well. Leaders, it seems, never stopped to think whether the end product would be truly educated or contributive. It was go for broke, *get there* before we run out of gas. It's the *system* and like it or not, it is deeply *nodulized* in all of us. "More-more-more—by any means possible! Just pass the "friggin' test"! The end will justify the means" becomes the growing mindset—a mindset that allows the invasion of countries, implicit changes in the Constitution, torture and the creation of a cloak-and-dagger police state.

Unfortunately, trying to solve the educational problem with a *'con'* approach involving piles of money or test statistics has always been doomed and always creates more problems than it solves. The recent sudden influx in bi-lingual populations only served to magnify the failure of the 'Ponzi Ed' system. The new waves of children (progeny of immigrants) are receiving the promised *profit yield* but the investors (immigrant parents) are not paying for the next *revenue boon*. The collapsing system becomes obvious and the new immigrants—NOT the tired, worn out, failed educational schemes—are blamed.

IF YOU MEET BUDDHA—(at the very least—run like hell)

People ask me, "What's your favorite educational model? Are you a traditional "three R's type of guy? A charter-school guy? A home-school guy? Are you *waiting for Superman*?" My answer is the same one I answer when people ask me about religion. My favorite system is innovation, freshness with an ever-increasing portion of curiosity. As Bertrand Russell said years ago: "The <u>will to believe</u> is not needed . . . but rather the <u>will to find out</u>, which is the exact opposite." To find out! Because when it comes to the best system for any given generation of people—I don't know and neither do you! But if we keep ourselves open minded and work to free ourselves of *nodulized* beliefs about habitual, time-honored '*rights*' and '*entitlements*'—we can continue to find out. Whatever the answer might be, it is certainly not the traditional educational belief system we now have in place. We may not know what the complete answer is—but we certainly know what the answer is NOT! It is very clear that what has gone before is NOT working in the twenty-first century.

DC economic professionals are fond of saying "a *vision without funding is a hallucination*." That is exactly the case with those who talk about educational reform. Those who say the answer is a minor change of this or that; put a private school here or there, really do not know they are hallucinating—pretending to be the knowledgeable Buddha, without being *'awake'*. Anything other than a modification of our thought processes and a total system change is another *hope*-filled hallucination.

POSSIBLE MODELS TO BEGIN CHANGE

1. Age requirements. In most states you must be 21 years old to drink alcohol. How about having a law that stated you must be under the age of thirty-five to serve on a school board or be a school administrator (plus the other general requirements). At first glance that may seem absurd, but let me explain why I think it is crucial to have a new educational plan that rapidly moves away from and beyond the old guard now running our failed educational system.

Usually, at some juncture during adolescence most of us have a series of moments, when an epiphany hits and we become attentive to a world larger than ourselves. We become aware of beliefs and biases, of attitudes and attributes that have shaped us; in short we become aware of our *nodules*. Interestingly, these *nodulized* beliefs and attitudes tend to persist into and through adulthood. (Of course 'there are exceptions'. You may be one.) But generally speaking, it is why the music—as well as the level of *electronic comfort*—acquired during pre-adulthood, *sticks*. True, you may modify slightly by taking a computer course or two; you may even become a quasi-musician but when the mid-life crisis strikes, it is the music and the computer/electronic skills acquired during your *era of awakening* that are most comforting and effective.

The point is that by the time most of us hit twenty years of age our societal-view is essentially fixed. The history that came before twenty is taken for granted; it is considered natural, ordinary and *real* while large portions of what comes after are viewed with suspicion, vexation or just plain revulsion. Parents of the forties abhorred *Boogie Woogie*. Parents of sixties were shocked senseless by Rock and Roll[4] just as parents of later generations resisted Funk, Punk, Rap and Hip/hop.

Given a reasonable peer group and economic solvency most of us hold on to our *nodulized* beliefs and attitudes about what we think is the *natural, ordinary and real* as we mature, taking in stride the strange changes that stimulate our suspicions and paranoid outbursts. (The world is going to hell in a hand basket!) Yep, as you age, if you are honest, you will admit your ever-increasing desire for the *good ol' days* when life was easy, simple and fun. The music was *'real'* music! You listen to *that* music when you have a choice. (And yes, YOU may be the exception.)

Whenever I run into a serious and articulate educator who is absolutely clear on how to solve the education fiasco, I check to see how his or her solutions fit into the popular societal-view when they were in their late teens and early twenties and sure enough, that's what usually fits pretty well as their answer. For the most part we can't help it. Those were the answers that helped shape us, they were the answers that held us in check and made us who we are. They *nodulized* us, how could anything be better? Answer: I don't know and neither do you! I do know that teachers who are closer to their students—who seem to have a sense of understanding (empathy?)—are also, *generally speaking*, <u>closer in age</u>. (You may be or know an exception.) In this contest it seems obvious that school board officials and those who are responsible for future policy should be in touch with what is natural and ordinary in today's societal scene.

What is the current cultural contour? This is key to keeping today's students motivated and curious about learning. Today's kids are not interested adopting the mores or world-view of your adolescence or mine, they want to create their own just as you and I did. If educational policy makers were disallowed after the age of thirty or thirty-five, the current educational system would certainly be closer to reflecting <u>today's</u> world-view.

Moreover, when there is a rise in adolescent connectivity (Internet, Texting, My Space, Face book, Twitter etc. etc.) a new language will emerge, which will influence the way younger human beings evolve thinking. No doubt Gutenberg's printing press—because of the immutability and seeming logic of the printed word—signaled the growth of a more disciplined and complex form of thinking. However, now that we have *texting* and *tweeting* and a new abbreviated *acronym language*, our ever developing thought processes, no doubt, will shrink to more personal, yet revealing, symbols. (Perhaps this is where telepathy or something close to it will emerge.) Not even our—by comparison—'long-winded' Emails can sustain the barrage of this new minimalistic thought process. At the same time there is also a rising tendency toward personal (rebellious?) expression and joining groups, gangs and new clans. Joining becomes a way to maintain an identity. Just as in the macrocosm (world at large) we see that globalization's most foreboding dynamic is the loss of National identity, the same dynamic exists in the microcosm. Tom Barnett[5] a brilliant world analyst has repeatedly pointed at *identity crisis* as one reason for the recent religious interest in developing countries, "when there is a heightened connectivity between societies . . . we tend to see a rise in church membership in the less advanced societies as individuals reach for religion as a way to maintain their collective cultural identity." When a society or a generation of school children face transformation, and/or even extinction, because of exposure to globalization, distressful crisis and anxiety are generated—big time! (Surely, most of us recognize adolescence as an era of *identity crisis* if nothing else!)

2. <u>New world capitalism</u>. Historically, capitalism as developed in America, was the envy of the world, primarily because in most fields it connected the dots between incentive, innovation and motivation. <u>Today</u> I write these words on a MAC computer and <u>today</u> Microsoft is spending millions on ads and inventive innovation to recapture the *geeks* who deserted Bill's *vista* (pun intended!). When competition becomes stifled and the capacity for innovation is *'unionized'* out of existence, the triumph of the capitalistic principle ceases. Motivation is minimized. Society at large pays a terrible price. This is precisely what has happen in a significant section of America's educational system. Instead of capitalistic incentives and competition, our educational system evolved into a baby-sitting bureaucracy by the demands of war and unions. When Dad went off to battle and Mom went to work, the school became the natural alternative for childcare.

Theoretically, the most efficient way of developing new educational systems has always been through private enterprise. But our major educational organizations have not been encouraged or pressured by the forces of modern market capitalism—with its inherent major and minor systems of competitiveness—to give anything but lip service to the development of new educational systems. Quite the opposite, government subsidies were handed to the new baby-sitting facility, and this *idea* became *nodulized* and persisted after the need was long gone. Babysitting in public schools at public expense evolved into an entitlement. Teaching to pass a test became the model where scads of money and teachers energy were distributed. *No child left behind* = "let'em in get'em out!" Qualified or not they are entitled!

While capitalism around the world relentlessly evolved—creating more efficient and innovative methodologies elsewhere—*education in America* became a *lame duck* and wallowed in confusion, stuck in what now amounts to Dark Age *uber-belief systems*. If this occurred in some small corporation, it probably wouldn't matter very much. But when it occurs in our most vital industry, which is without a doubt the <u>education of our next generation</u>, it matters more than we can possibly grasp.

Some experts believe it is the government's responsibility, as the guardian of America's survival, to become the catalyst for educational transformation.[6] But, the experts also say, this cannot be just a matter throwing more money at a decaying system.

3. <u>Educate potential parents</u> (no matter how young or inept). This would not be popular and would definitely require money and whole new approach. Why? First of all, because rat pups that are licked and groomed by rat moms grow more brain cells! What? (Yes, there is a point to that fact) Secondly, at the University of Minnesota they studied the '*mother /child attachment patterns*' of 42-month-old *pre-schoolers* (we are talking

babies here!) and discovered they could predict with a 72% accuracy which of these pre-school babies would graduate from high school. This BMA—Considering High School graduation rates in all the large urban communities of America are at an all time low. Doesn't it seem logical that some time and money given to "mom training" could reap some remarkable results? GS or BS? You be the judge.

4. <u>Demand accountability</u> and a new kind of attitude. President Obama was not far from this idea when he suggested in one of his quasi 'State of the Union addresses' that high school drop-outs were not "just letting themselves down . . . they were letting their country down." Imagine being labeled a traitor to your country if you are slacking in study hall. Unthinkable? Maybe it is time to begin thinking the unthinkable! (Traitors have been hanged! Talk about motivation!)

A teacher friend of mine quit teaching because the "kids today don't respect anything—including patriotism." And it seems true—disrespect, rather than accountability, is being woven into the fabric of modern culture. In part because altogether too often our great heroes are exposed as cheats, frauds, liars and drug users. Priests are exposed as child molesters; Olympic champions are exposed as steroid abusers; Bankers, Automaker executives and Union officials, whose greed collapsed the National financial system, are now unabashedly requesting handouts from school children (albeit two generations from now). Everyday, web search engines reveal the cute curbed video streams and bi-lines, where public figures are seen being arrested, assaulted, or exposed. Infidelity and fraud are fodder for standup comedians. Little wonder there is so little respect for teachers or any authority figure. Our entire system is obsessed with the crass and the rude and it must change or cease to survive.

5. Provide incentives. Jefferson served it up as follows: "And say, finally, whether peace is best preserved by giving energy to the government or information to the people . . . Educate and inform the whole mass of the people. Enable them to see that it is in their interest to preserve peace and order, and they will preserve them. And it requires no very high degree of education to convince them of this. They are the only sure reliance for the preservation of our liberty." (Thomas Jefferson to James Madison, 1787.)[7]

It is no longer just a matter of the carrot or the stick. A critical mass has accumulated and, in this brief moment in time, it has become a crisis and it is now a matter of creating a new motivation in our educational systems. The words *"it is in their interest"* seem to lack the idealized beneficence of a free handout from an altruistic government. Rather, in this concept we are talking raw competition and incentives for success. A really new educational system would need to compete for consumers—innovation would follow. This has not been the case for decades in our public education systems. Why not? Historically our *free educational system* has existed under the premise that IT was so important to America that it should be treated differently from other Capitalistic competitive enterprises. (Too big to fail!) As a consequence massive tax supplements have left these decaying educational systems free to ignore the vital business maxims that allow other American capitalistic systems to thrive. Namely, new technologies replace existing ones because they are cheaper, more efficient and more consumer friendly. Our educational enterprise has been free to disregard the need for '*incentive*' in the market place, and instead, has been given wheelbarrows full of taxpayer money with which it could continue to prop up its concretized, centralized, ineffective, outmoded systems. Without market pressure to innovate or find alternative ways to instruct, the educational system provided the American public with an inferior product. Objections were few and far between; after all it was and still is free babysitting!

As we have recently discovered, Toyota took the market share away from Ford and GM by making innovative hybrids while the gluttonous US companies still turned out hulking SUVs. Not only were the US companies not punished by the marketplace, they were rewarded by the so-called *bailout*. But, turning a blind-eye to cynicism, we hopefully have oversight and these corporations are *hopefully* accountable. We now *hope* they can no longer ignore the market driven changes that have so dramatically made American capitalism the envy of the world. Yet, we have been bailing out the educational system for decades. It failed repeatedly and we just continued to throw more money at it! The auto industry in America has nothing on the educational industry. The parallel is not difficult; in a modern world of intellectual shortages, we can no longer afford the luxury of allowing old-fashioned, non-innovative, non-competitive programming to be at the heart of our educational system. Social evolution continues unabated, whether we like it or not, society changes. Prior to our 'Ponzi Ed. scheme we were a society built on the *value* of graduates. We have become an ignorant nation based on the *volume* of graduates. As a consequence we have fallen behind almost the entire industrialized world. Our *educated graduates* are not even on par with many barely developing countries.[8]

A NEW KIND OF MOTIVATION

Please pardon me if all this sounds a bit contemptuous—perhaps *despairing* would be a better perspective. Even in today's anti-climatic environment of "The Audacity of Hope", I seriously doubt that without a complete overhaul of our entire system, anything but *more of the same* will occur. Certainly, most of us have heard the positive thinkers say we are in the midst of a *revolution of consciousness*. Whenever I hear that I fall silent with disbelief. How can any thinking person talk about a *revolution* where there is no pain, soreness or grief? Have you ever seen or heard of such a thing? In 1776 there was a painful American revolution, and another in 1860 where brother fought brother to free slaves. It took 72 years of painful revolution for the 19th amendment to take hold and women to win the vote in 1920. What are the thought processes of politicians and ensconced educators that lead them to think they can have change in the educational system without casualties, where no one has to give up anything? What some of them describe as *change* sounds more like a merrymaking frat bash than a revolution. Yes, all over this land of ours we are having full-on celebrations and in the words of journalist Tom Friedman ". . . it is mostly a costume party. It's all about a certain *look* . . ." (a certain way and mouthing certain liberal or conservative buzz words.) "Yes, step right up ladies and gentlemen, in (this) revolution everyone gets to play, everybody's a winner, nobody gets hurt, and nobody has to do anything hard . . . That's not the definition of a revolution. That's the definition of a party".[9]

By now it seems obvious that *'hope'* won't cut it! *Positive words* won't cut it! *Business as usual* won't cut it. Without a revolution: a new intelligent re-design of the whole system, a response to market demands rather than handouts, and the instilling of a fiscal <u>ethic</u>,[10] (undoubtedly hurtful to large segments of affluent citizens), we will not have a chance-in-hell of turning a vision into reality. This will not be *easy,* and, in the context of educational revolution, the word *hope* should never be allowed in the same sentence with *change*. The core of this *change* insists on preserving and restoring America's most valuable resources: educational innovation and intelligence-based progress within an ever-changing system. These resources are rapidly being depleted by current *status quo attitudes* and it must stop—now!

This kind of change will also insist on breaking our collective addiction to mediocrity and our mass worship of the ordinary. This is not something we do on the side while we work on the *next big thing* or figure out how to *dumb-down* the world news so that it will be digestible in entertaining sound bytes. Legitimate news is defined as information dissemination NOT entertainment.

The words *easy* and *comfortable* cannot be a part of this revolution and should not be accepted. We need leaders who will admit openly that this is hard; it is the most difficult and most vital undertaking ever—including every war we have ever fought.

SUMMARIZING

This crisis in education is affecting everyone. It is not a matter of the *haves* and the *have-nots*. It is not a matter of urban vs. suburb or any other of the population divisions. It is a matter of survival for all of us! How to make this known? America won WWII because of a *painful* combined effort. No single military general or single battle or even the *bomb* was responsible. It took the sacrificing of an entire generation. If I were to name a single motivating force behind the day-by-day history making effort, it would have to be the propaganda (*media at large*) that always insinuated that if America was to survive, its citizens must realize they were *in it* together. Yes, news information was undoubtedly bleak at times but *the greatest generation* listened and learned. Furthermore the flow of information was repetitive, constant, no slacking—no let up. It was a prime example of *Neuro-nastics!* The movies, the newsreels, the radio, the magazines and printed page all connected to repeatedly inform and make possible the greatest united effort ever witnessed on this planet. American *Media* facilitated *the greatest generation's* survival because *it*, more than any other single format, reiterated, recapped and repeated information that kept Americans informed and educated about what was needed and when.

Once again, today, in this crisis of crises, the *Media* needs to step up to unite and harness its full potential to *educate* the people. Stop the '*dumbing down*' and the '*numbing down*'. Stop indulging the mass worship of the ordinary. Stop providing a public pulpit for cheats, frauds and liars. Yes, at first this may impinge on sales, but is time for the *Media* too, to sacrifice. Once again, it is a matter of survival.

Secondly, stop trying to re-vive a failed system. Instead, begin anew. Get serious about the real challenge to our survival. Yes, it may be true that no one has yet come up with the answer. But we all know what the answer is NOT! The greatest challenge confronting this generation isn't faltering ice-shelves or too much heat in the world; rather, it is too little heat in educational leaders—too much ice in their hearts. Change? Not on their watch. Let the ship sink, tenure must remain sacrosanct no matter how incompetent or unmotivated the teacher. Let the American dream fade; so what if only 50% of kids in Los Angeles finish high school, that can't be helped. Yes it can! Start again, from the ground up; rebuild a whole new kind of system.

It seems obvious that the *sacrosanct* Jeffersonian educational model of government that sponsored free babysitting is no longer working. Like so many other worn out but thoroughly entrenched (*nodulized*) bureaucracies, it needs to go! Over the years we have all watched as the same dynamic was repeated. When the power or salary of an entrenched educational '*monarch*' was threatened, the ripple of counter-measures was stunning. Educational bureaucrats have well-established identities and the evidence reveals that they .would rather see the death of their charge than to change their mind-set, just as abusers would kill their victims rather than allow them to escape. Anyone who has ever seen or been in an abusive relationship knows that to leave can be life threatening. Hitler tried to take down all of Germany when he realized the war was lost. Only blatant disobedience on the part of a few staff generals prevented him. If the race to nuclear power had been won by the Nazis there is little doubt they would have used it destroy the world just as religious extremists threaten today. If an abuser cannot dominate and control a thing, then they will not allow it to continue.

However, the more immediate threat to America is not nuclear or economic, although entrenched governmental mind-sets (also called habituated *brain nodules*) may well produce both of these catastrophes or worse. No, the most imminent threat to America today is the absence of an educated next generation. This current generation can become the new "greatest generation"[11] by providing the next one with a superlative education. Sure we'll undoubtedly make mistakes, but in the name of sanity we know the current system does NOT work! As Leonard Cohen implores, "Start again, . . . forget your perfect offering, there is a crack in everything . . . that's how the light gets in . . . that's how the light gets in . . . ."[12]

CHAPTER 11

"WAR" & OTHER HEROIC *NODULES*

"The Dove is bought & sold & caught again . . ."

For more than fifty years the voting citizens of the US of A have been ignoring a bit of presidential advice—perhaps because it was too frightening to be taken seriously. The hero of WWII, Dwight D. Eisenhower gave some alarming counsel in his farewell presidential address in 1961: "*In the councils of government, we must guard against the acquisition of unwarranted influence, by the <u>military industrial complex</u>. The potential for the disastrous rise of misplaced power exists and will persist*".[1] Ike's warning against the military *duping,* which he foretold would manipulate us, (Read: '*Scare hell out of us*' and, no doubt, create brain *nodule* changes.) has largely gone unheeded. We seem to take it as a matter of course to spend obscene amounts of tax and borrowed dollars on defensive war research and killing products. Question: We Americans have followed this destructive path for so long our brains have undoubtedly been transformed (*nodulized*)—can we now do otherwise?

Historically, the next president in line (a chic and stylish PT boat hero) J.F. Kennedy, just three days after Ike's palpable warning, vowed to "*pay any price, bear any influence*" of the (military-industrial complex burden), "*meet any hardship*" in defense of *freedom*. That particular president held a whole generation in the grip of iconic imagery; so much so, that youthful generation seemed to believe that JFK was their political *messiah*. Whatever he suggested was taken as gospel and since then we have been subjected to almost unbounded foreign commitments, mostly involving showing off our *military might* and our expensive war toy superiority. These awesome killing machines have held us '*dupidly enthralled*' with our military supremacy for a half century. Exactly as Ike predicted, fear mongering and monetary waste have brought America (us, you and me) to bankruptcy. Because of war, every man woman and child in America today is mortgaged to the tune of $175,000. We spend six times more than any other nation on earth (700 billion) via the MIC (military industrial complex). Every attempt to slice and dice this gluttonous behemoth is countered with *bribery*—our elected politicians have sanitized the word 'bribery' and now call it *lobbying*.

And since January of 2010 our United States Supreme Court has officially legalized *'bribery'*. They overturned previous spending limits and currently allow unlimited corporate cash to literally *buy political power*. Everything the American public hates about politicians *'on the take'* is now legal. This decision represents one of the single greatest demonstrations of contempt for legal precedent in the history of democratic America. It will now be completely up to our individual politicians to decide just how much is too much. Who believes they will? LOL

Political leaders, who accept the *lobbying,* justify the MIC by appealing to popular passions (fear) and prejudices (paranoia) and the brain *nodules* continue their synchronous march onward creating even more propaganda by their very rhythm. Listen to the beat: "We must not render our nation defenseless" . . . "they are just waiting for us to let our guard down." Dum de dum de dum de dum. Dumb!!

We need more faux right-wing thinking comedians like Stephen Colbert, who publicized a political rally to *'Keep Fear Alive'*. The title alone "captures an entire war strategy . . ." Not only does this kind of jibbing satire provide a reality check, it also provides a badly needed prompt to laugh out loud at <u>all</u> the non-sense related to a war to end war. Even the backward village idiot would acknowledge that ultimately war is a 'lose/lose' activity for *'everyday man'*. Yet, Winston Churchill once said, "The story of humanity *is* war followed by precarious interludes of peace." He was unapologetic in observing that our human history is one of almost continuous murderous strife. In fact, thousands of books are dedicated to describing the sources and nuances of war.

Is any war, including WWII a good war? Did fighting around the globe help anyone? Or stop future wars? Nicholson Baker in *Human Smoke* answers with an unequivocal NO! By the way, if you are still pondering the questions regarding the efficaciousness of some "righteous" war, I challenge you to read his book and come away feeling the same way. Still, war continues to be trendy and even heroically chic—new generation after new generation—notwithstanding protests and bankruptcy. Then in a moment of lucid clarity we ask: Since religious values have always dominated our American political/civil rhetoric, how is it that we so quickly accept the MIC and killing other human beings as a solution to our predicaments? How did America's religious ideals and warring become so inextricably bound together? You want justification? *Surely wars to deliver American democracy and technology to undeveloped backward counties will prove that the ends justify the means*—won't they?

REASONABLE REASONS FOR WAR?

Typically young people are magnets for the propaganda of adventure. The deep baritone voice says, "Be all that you can be" with a vivid visual background of technology and futuristic war-toys. These kinds of half-truth cants are very appealing to the kids who have heroic dreams. Especially for those still desirous of/or needing some sort of external authority to provide direction for their lives. The dramatization of war on TV, repeated heroic stories, coupled with two or three months of tough 24/7 basic training can grow some pretty significant *nodules.*

Talk about *Neuro-nastics*! You better believe that by the time a youngster is finished with the basic *'training'*, a bona fide n*odulized/neuro-nastizied* <u>True Believer</u> has emerged—or his/her discharge is immanent!

MOTIVES FOR JOINING

1. The *'floundering syndrome'*. It is well known that the age between 17 and 24 is a second *tweener stage*. The first being the age of 10 to 13 when children are thrashing about in an effort to become authentic *teenage kids*. The *flounders* are kids battering on the door of adulthood and being turned away. Jumping into college is one option, gangs and drugs are another and the armed forces represent a third.

College requires a high school diploma (which stats say only 50% of kids from Los Angeles public schools acquire) and across the country the percent of students earning a standard diploma in four years is shifting downward with each passing year (e.g. 69.2 percent in 2006 to 68.8 percent in 2007, according to an analysis of recent data in Diplomas Count 2010). It was the second consecutive year of decline, says the report, which was released by Education Week and the Editorial Projects in Education (EPE) Research Center, a nonprofit in Bethesda, Md. Racial and ethnic gaps persist, the report notes. Forty-six percent of black students, 44 percent of Latinos, and 49 percent of Native Americans did <u>not</u> earn a diploma in four years. And the TV blares on "be all that you can be!" Those can be attractive words to a *floundering kid*.

2. Another reason the <u>war</u> construct holds some young adults in its grip is that when they hear about soldiers volunteering to return to the battlefield in spite of having just returned, they imagine highly motivated and excited warriors and fantasize about their prospects of participating in some heroic act in combat. Most fail to recognize that for many of our young aspiring adults it is not just the love of adventure but the love that soldiers have for each other that keeps them coming back to the killing fields. TMI? *Think about it*! You've heard guys say they going back for a second and even a third deployment because the "guys over there are

my family and they are depending on me to watch their back." They talk about a kind of closeness and a family-like dependency that they have rarely, if ever, felt before. If you listen carefully it sounds very much like a description of a high-quality marriage. Please! We're not talking '*gay*' here, what these guys are describing is something they have longed for since infancy. To be cared about is to care. This is reciprocity in its basic form. One of the first long *bog* words I learned pre-kindergarten was *reciprocity* (that for every action there is a reaction). My mother was oft heard saying, "There must be reciprocity in all things." Yes, at first the word *bogged* me down and while it took me some time, I did finally learn.

But reciprocity has a dark side. It is also about making other words (and ideas) from the original words and ideas; as well as what 'gravity' could do and being lied to, and inhaling exhaling and inhaling and exhaling again—yin /yang. War has its reciprocal dimension, not just national debt but the other side of heroism. It is the price we pay in the devastation of young minds and bodies and it is almost too staggering to talk about. Even with basic family love at its core, after the second or third tour, Post Traumatic Stress Syndrome is no joke!

3. When chic media personalities offer their recurring praise for the returning Iraq or Afghanistan *hero* they become the motivational PR billboard to sign up the

next *tweener*. He/she is asking, "What shall I do with my life?" The answer: Volunteer to join up and "be all that you can be". Yes, you too can become a national '*hero*'; just volunteer, go to basic training to grow a new 'war hero *nodule*' and "learn to be all that you can be." Learn to kill so you can go over there and keep the peace. Is there an oxymoronic statement in there someplace?

Perhaps some countries really do need to keep a war going so the listless "lost *tweeners*" within will have <u>purpose</u>. It certainly seems to be the case for the various countries around the world that allow *terrorist organizations* to thrive within their borders. What will their *tweeners* do without the performance of an act of martyrdom and promise of imminent paradise? "*Strap on the bomb kid . . . 'be all that you an be!'*"

A CHANGE IN CULTURAL CONSCIOUSNESS

It appears that the *liberal-secular Europeans* have lost their taste and talent for war. Some say that WWI and WWII formed a kind of one-two punch that knocked the *love of killing* out of the continent.[2] Even the briefest glance at the history of Europe reveals that historically their wars have been the subtle incubator of idealized heroes and irrational leaders. But today in Europe there appears to be a totally different context for an anti-war renaissance. Apparently, war is now seen—more or less—as something like an offense or even crime to be resisted and overcome—even among the college age youth. Actually, a much-touted evolving shift regarding war in the new European union appears to be creating a different kind of ethical calculus within the European culture. It now appears that Nations of the Continent have no desire to become ultimate military powers. They want to trade, not fight, with their neighbors. They are certainly no longer considered as *superpower puppets*. Unless the European peoples would have a dramatic change of *heart and mind* and chose to give up their emerging anti-war identities they will more than likely continue to evolve a warless advanced type of society.

Since peace has so many advantages, what would motivate a change in their new identity? Possibly a messianic demagogue or a resurgence of some religious fanaticism could account for such a backward step. However, that likelihood in such a liberal/intellectual climate is highly improbable. In fact, almost daily news headlines declare their independence from *western ideological influence*.

It appears that the rest of the so-called civilized world, especially American politicians, needs to take a long look at Europe and take a long war time-out. The world desperately needs an extended resting time to give some other nations a chance to evolve beyond primitive excuses for war. As long as ANY justification for war is accepted, *evolving-humanity* will be stuck repeating the cycle of primitive crime that goes by the name *justifiable war*.

WAR ALTERNATIVES

Actually, a national shift regarding war has already been noted in America. Not only are Americans sick of the longest war in their history; there is indeed a "different kind of moral calculus within America's world view". When China just *swooped* in and took over Tibet in a bloodless coup, not a significant word of international protest was noted.[4] By now it seems obvious that Nations of the world including America, will not get involved in combat with major powers—certainly not superpower puppets who inhabit the United Nations.

However, what most of the Western world does not seem to understand is that what Islam failed to accomplish in previous centuries, it can do without a war by immigrating to countries that tout their progressive 'openness' and their willingness to celebrate cultural differences. In recent times tens of thousands of Islamic believers have moved, immigrating to 'open-minded' countries like France, Germany and Canada. And then since one *interpretation* of their religion almost insists that a devout Muslim have more than one wife—as many as four is permitted—each wife may then have four or five children. The end result is to turn a Christian or agnostic majority into a minority. This is precisely what happened in North Africa and is now happening in the Netherlands, France, Germany and the UK where citizens are justifiably worried that their nation's character will transform fundamentally as Islam grows within. After all, in Muslim dominated Eastern countries like Afghanistan, "gender issues" are relegated to the back burner or at least considered as "negotiable issues". (E.g. Like allowing girls to attend school without the fear of having acid thrown in their faces, or allowing women to safely walk the streets without a male relative in attendance.) We're talking about allowing the female gender to be fully human. These issues should not be negotiable. But as the Islamic influence increases in "progressive countries" they will make such demands via exercising their right to vote in their newly adopted democratic countries.

The freethinking intellectuals of *open progressive nations* have pretty much eliminated religious ritual from the public politic. But when a religious majority begins to demand their 'rights', we will see the beginning of another *bloodless take over*. Holland and France appear to be contemporary examples.

For the most part, *'open secular countries'* are fearful of being accused of being intolerant, or "Islamophobic". The consequence is that the politicians, as well as the PC populace, back down and do nothing while their country and culture is invaded and subverted by immigrants who since they have no desire to *'fit in'*, will *'out breed'* the nationals and dominate the country within a few generations. Since Christianity has faded in most of these countries and moderate secularists avoid the dirty business of politics and military intervention, they can only hope that 'modern common sense' will catch on within religiously fixated countries, particularly among the women, which for the most part have learned to *'believe'* rather than *'think'*.

UBER-MYTHS

1. <u>Immigration is harmless</u>. Earlier I suggested, "possibly a messianic demagogue or a resurgence of some religious idealism (fanaticism) could account for a backward step and that such a likelihood in such a liberal/intellectual climate is highly improbable." But this new kind of invasion—bloodless war—is highly probable and even now the inroads of this kind of incursion are becoming clear. Countries with a Democratic bent are particularly vulnerable to this new kind of *'bloodless'* war.

The rest of the world needs to take a long look at Europe and take a long time-out. Both war <u>and</u> immigration need a time out. Democratic nations desperately need a resting time. Other nations need a chance to evolve beyond primitive ideologies because we still observe that in bringing resolution to political disputes, revenge (tribal war) is still primary in most of the Mid Eastern world. As long as there remains ANY justification for having absolute power over others—having people or a gender considered subservient—evolving humanity will be stuck repeating the cycle of primitivism.

2. <u>Arms control is another uber-myth</u>. Any objective observer can easily see that it is utterly futile to institute rules and limitations for the killing devices of war. They are devised to KILL! America as well as other nations actually conducts, what is euphemistically called *'war games'* to test their newest killing machinery. How often we have seen freedom and individualism fly out the window in times of war. Marshal Law and authoritarianism becomes the norm. We then proceed to give-up-and-in to the monsters in the name of patriotism. Behind all of it is the *uber-belief* that somehow some war will "show them." Once again we can see the need for some extreme Neuro-nastics REHAB!

3. <u>All volunteer army</u>. America's all volunteer army has fought two wars while the vast majority of its citizens does NOT personally know a soldier or have a single relative involved. Had there been a national draft, American citizens wouldn't have tolerated a war. GW Bush didn't need bodies or in the short run money. Mercenaries (kids looking for adventure and high paid experts) volunteered to fight and strangers from overseas produced the finance. Like Ancient Rome and Victorian England, American politicians can

utter their high sounding rhetoric and continue to fight exploitative wars with paid mercenaries BUT ultimately this will destroy America financially and ethically.

Man's inhumanity to man is not just brutal, it is insanity personified. There is just no other word or way to describe it. We may try to *sane*-itize it by calling for "rules of engagement" or banning nuclear weapons or outlawing this land mine or that nerve gas but on some level we all know it is craziness beyond words. Trying to 'civilize' or humanize war is a psychotic delusion. Einstein said, "One does not make war less likely by formulating rules of warfare. The only rational advance for mankind is to outlaw war itself."

CHAPTER 12

ADDICTION *NODULES*

sex, drugs, videos & other fantasies

Fantasy and reality seldom meet or walk together on all fours—or even lay together for that matter. Often, when discussing sex, and more particularly sexual fantasy, the term *orgy* crops up. For guys it can sometimes mean ménage a trios' or a three-way sexual session with *himself* as the center king pin and two compliant women. Generally however, reality turns out as a less-than-exciting adventure—more like downright grueling. Two competing women, kissing, hugging and touching each other in the presence of a male voyeur is the romanticized fantasized landscape. But, which one was he more attentive to? With which one was he more tender or aggressive? Then comes the sudden awareness that he is being watched as much or more than he is watching. He becomes compulsively choosy in the selection of his erotic evocations and gestures. Like a fox, he finds himself being scrupulously *fair and balanced*! (Yes, that was a deliberate pun.) The *here and now* gives way to an attempt at evenhanded attention and improbable impartial consideration. Most of the time no matter how he tries, he fails—his *significant other* turns her head while the new paramour withdraws into feigned indifference.

Sexual fantasies—when acted out—often turn out to be like an initiation to prove ones worth or virility. A realistic analysis reveals a scene of addiction that feels more like the mythical Sisyphus rolling a colossal rock up the hill—without the relaxing stroll back down. Addiction is 'Having no choice' for the moment. Expending too much *oomph* trying to fulfill an *'unfulfillable'* fantasy is always, without exception, an invitation to addiction.

SOME OTHER ADDICTING FANTASIES:

1. **_Pornography_**: In 1869 a little moth was imported to America to jumpstart a silk industry. As usual, human good intentions went '180 degrees *south'*. The moth had an insatiable appetite for indigenous trees such as Oaks, and Maples. Numerous unsuccessful attempts were made to destroy this pest but for 150 years the Gypsy moth devastated American forests. A breakthrough came in the 1960s, when scientists noted that the male moth finds a female to mate by seeking out and

following her mild aroma, or *pheromone*, which is a kind of chemical scent capable of acting outside the body of the secretor to impact the behavior of the receiver. Scientists were able to mass-produce the pheromone and permeate the moth's environment with it. This unnaturally strong scent overpowered the females' normal ability to attract the male, and the excited but confused males were unable to find females. Trapping was also used by setting out Pheromone-infused traps from which the gentleman moths could not escape; the 'tumescent male' enters looking for a female, only to find himself with a fatal proxy instead.

What does this have to do with *pornography*? True, while there is no record of anyone ever dying from looking at porn, *pornography* is nevertheless is like a *visual pheromone,* (a powerful, $100 billion per year brain drug that is altering human sexuality by *'inhibiting orientation'* as well as *'disrupting pre-mating communication between men and women* by permeating the *Internet atmosphere*). We are in a losing battle against *pornography* because so many men and women, continue to believe 'Internet Porn' is neither a *drug* nor *addictive*.

At this juncture let's review a little of the science related to answering the question as to whether *pornography* is an addictive agent. i.e. a *drug*:

Epinephrine is a *drug. (More commonly known as* Adrenaline) It is used in surgery or in emergencies to jump-start the heart when it slows or even stops. The human body/brain makes epinephrine and it causes the heart to pound and race. It is the same epinephrine given by a surgeon or to those with super-allergies to honeybee stings. Perhaps <u>you</u> may have had this kind of experience (heart pounding) when broaching a sexual encounter? TFS (*Thanks For Sharing*—or NOT!)

Dopamine, a kissin' cousin to epinephrine, is also a drug. It is an excitatory neurotransmitter that tells the brain to tell the body (particularly certain parts: "*Go Dawg GO !*") Dopamine is specifically important to the parts of brain that excite us and tell us to *move and groove.* Of course, both *epinephrine* and *dopamine* are drugs in every sense of the word, regardless of where they are produced. Pertinent to the subject of porn, it happens that both of these brain drugs are key to human sexuality. In addition, besides its role in body movement, *dopamine* is a neurotransmitter in the pleasure/reward system of the brain. So some pharmaceutical companies use this drug in their prescriptive anti-depressants. Dopamine then, is used as a *ladder or step* up from and out of clinical depression.

In general, the reward system of the brain always involves some form of sexual excitation. (Imagine that!) As we discovered in earlier chapters, the outside lumpy/bumpy-like covering of the brain is called the *cerebral cortex.* It's *a thick gray/white layer of nodule-like nerve cells that facilitate associations and volitional thought.* In the front part, over the eyes, are the *frontal and prefrontal lobes,* which are vital to anticipation, judgment, and inhibition. (If the brain were a car, the frontal lobes would be part headlight and part brake pad.) These lobes also have important connections to the pleasure pathways, so pleasure can be facilitated or inhibited, as future planning and good judgment would determine.

Toward the middle of the brain is the *nucleus accumbens.* This almond-sized component is a key pleasure reward center, and when activated by neurotransmitters such as dopamine it creates an enhanced motivation for pleasure. Dopamine is essential for sexual desire (libido) to even exist. Without it, we would not be as motivated to survive: e.g. to eat, procreate, or even allow our eyes to take in a cute derriere. TMI?

However, and this is critical, <u>*the overuse of dopamine causes addiction.*</u> i.e. When overused, a downgrading occurs that actually diminishes the amount of dopamine available, and the dopamine cells themselves start to atrophy. The *'reward cells'* in the nucleus accumbens are then famished for dopamine and exist in a kind of state of constant craving. This results in a recalculating of the *'pleasure thermostat'* and a *'new normal'* results. <u>This IS the addictive state</u>! The person must now act to boost the dopamine to levels—just to feel normal.

As with any addiction, stronger and stronger stimuli are needed to produce the dopamine. In the case of narcotic addiction, the addicted person must increase the amount of the drug to get the same high.

In pornography addiction, research has shown that progressively more stimulating images are required.

As the addiction continues, the frontal lobes also actually shrink. It is kinda like turning off your car *headlights at night* and keeping your foot on the gas and the brake at the same time—*wearing out the brakes*. This physical and functional decline in the judgment center of the brain causes further impairment to the ability to anticipate and process the consequences of the addiction. This condition has a rather frightening name: *hypo frontal dysplasia!* And please note a similarity in the behavior of patients with traumatic frontal brain damage and those with porn addiction has been recorded.

Frontal-Lobe-damaged individuals exhibit four symptoms: 1. They are *impulsive*, i.e. engage in activities with little or no concern for consequences. 2. They are *compulsive*; i.e. become fixated on particular objects or behaviors, and *have* to *have* or *do* them. 3. Their *affect* becomes *labile.* (e.g. unpredictable mood swings.) 4. They repeatedly show *impaired judgment.*

Hypo-frontal dysplasia, or shrinkage of the frontal lobes can result from brain trauma or from pornographic addiction. TMI?

SOME COMPARISONS:

In the *Journal of Neuroscience* several studies of *cocaine addiction* showed shrinkage in several areas of the brain, particularly the *frontal lobes*. A 2004 study showed similar results for methamphetamine. Not surprised? IDTS. We expect brain damage as a consequence of drug use. But in a study about a *so-called* natural addiction, (e.g. *'overeating'*) Dr, GJ Wang, in the *Proceedings of the National Academy of Sciences* 2006, showed frontal lobe shrinkage was very similar to that found in cocaine and methamphetamine addiction. Recently (2011) Ashley Gearhardt published a study done at Yale University suggesting that eating addictions and cocaine addictions are similar in MRI brain photos. Women with acknowledged eating problems had their brains scanned while they looked at a picture of a milk shake covered with chocolate syrup. Brain areas related to pleasure and craving lit up much the same as they do when drug addicts are anticipating a fix. When the women were allowed to sample the shake they showed les-than-normal activity in the reward center, meaning they felt less satisfied. They had developed a need and a tolerance for sugar and fat. Like drug addicts when they get what they want there is less fulfillment than expected so they want more. Another study published in 2007 of persons exhibiting severe sexual addiction produced almost identical results to the cocaine, methamphetamine, and food addiction studies. As suggested in research throughout this book, the good news is that we can reform physical brain tissue changes. After REHAB *Neuro-nastics,* a return to near normal frontal lobe mass was shown with both methamphetamine d*rug addiction* and eating *addiction.*

ADDICTION IS ADDICTION

Brain scans of all types (fMRI, PET, and SPECT) show frontal lobe damage with cocaine addiction, pathologic gambling and overeating. As stated in the journal *Science*, "as far as the brain is concerned, a reward's a reward, regardless of whether it comes from a chemical or an experience."

Dr. Eric Nestler, head of neuroscience research at Mount Cedar Sinai in New York (one of the most respected addiction scientists in the world), published a paper in the journal *Nature Neuroscience* in 2005 entitled *'Is there a common pathway for addiction?'* In this paper he said that the "dopamine reward systems mediate not only drug addiction, but also *natural addictions* such as pathological overeating, pathological gambling, and sexual addictions." This is science—not just a straight-laced POV.

2. Polygamy

Blowing your own horn is a familiar theme in men's locker rooms. (no pun intended) Here I'd like to consider some other kinds of addictions. For example, what do descriptive orgy scenes, multiple spouses and swaggering men's clubs have in common? More often than not, men boast of their sexual prowess, primarily to other men. More often than not, such boasts are a cover up for feelings of inadequacy. Think about it, a man is indeed virile if he can participate in coitus two or three times a night. As he passes the age of thirty, the number reduces. By age fifty most men have drastically reduced their spousal sexual encounters to a weekly venue—if their wives are lucky. Multiple orgasms, and multiple '*satisfied lovers*', as a general rule, are the province of the female gender. Quite naturally a woman can easily satisfy many lovers without fantasizing anything. Most men, on the other hand, find it difficult to keep one woman sexually satisfied over an extended period. A polygamous man with two or three wives in Utah (or a four wife harem in the Muslim world) may provide public perception that he is virile (*either sexually or monetarily*). But, in the final analysis, most are addicted to putting on a public demonstration of their supposed sexual prowess to satisfy the sexual fantasies of other men. It's a show for so called *lesser-qualified men* who either feel inadequate or look upon the display as something to seek and imitate.

Furthermore, most of the time, the polygamist keeps his wives pregnant. Thus he can not only be relieved of responsibility to provide frequent sexual satisfaction but he can convince other men that multiple offspring is another idealized fantasy. Over-population becomes the next fantasized paradigm to be emulated by other men. JJ. If the show of money (another addiction) is at the center of the polygamous display then power and possession become the exhibits. Again to his own opprobrium the polygamist is saying to other men, "See how powerful I am . . . I can literally <u>own</u> these women." When asked about satisfaction, few women who are in polygamous relationships will gainsay the party line. When asked, "Did they feel fairly treated, and was he able to provide sexually for all of them equally?" They are unable to give solid answers—other than "I follow God's will . . ." TFS (*thanks for sharing*—nothing*)*. While, in many cases, the husband is able to provide a comfortable home for them, there were also many cues and clues that polygamous communal living was far from a halcyon ideal.

SCRIPTURAL AUTHORITY FOR POLYGAMY:

While the Hebrew Bible appears to condone polygamy by the example of some of its Patriarchs, the Christian Bible (new testament) is silent on the topic. The Quran[1] however is pretty clear on the subject: It says that "if a man fears he is unable to be fair and just to his wives, then he must only marry one". As a matter of fact, when a Muslim male population was decimated via barbarous intra-tribal wars, it seemed necessary for those surviving males to take on the responsibility of the widowed women <u>and</u> their children. The inspired counsel, as elaborated in this chapter's *endnote* number one, *responsibility,* <u>not</u> sexuality, was the motive for taking an additional wife WITH her children. Read in its context, this is what the Prophets' benevolent Quran message was about. He was definitely <u>not</u> talking about taking on a virgin teen-ager as a wife. Like so many '*inspired writings*' they can be twisted to suit the interpretations of mullahs and *perps* alike.

3. Pie in the sky:

Because our ability to imagine a future and fantasize about it, our addiction to future abstractions has probably been around as long as *consciousness* itself. In fact, we sometimes think of navigating the future as some kind of ultimate mental gift. It certainly has facilitated a whole lot of anxiety and anxiety reduction. Imagining eternal '*well being*' for our loved ones and ourselves and then inventing ritualized means for achieving that imagined '*well being*'; that invention could also be considered our greatest gift. The "pie in the sky" rituals certainly have comforting value as well as addictive qualities.

It is *memory,* however, that keeps bringing us back to our genuine but transitory *most valuable gift*—the '*here and now*'. Our memory of yesterday is in verity, only one of many cortical engrams and our anticipation of tomorrow but a pre-frontal dream but every today well lived makes every yesterday a memory of contentment and every tomorrow a possibility of another *today*. Thus '*here and now*' is really all there ever has been and all there ever will be. And, FYI, short-term *memory* makes all that possible.

The idea that we currently have a solid grasp of the power and potential of our neuro system is pure foolishness. If you need absolute-finished answers you won't find them here. We've only barely scratched the surface. Mind and thinking dilemmas that seemed intolerably mysterious a decade ago are seen today as amenable to open discussion and discovery via new technologically advanced instrumentation. This kind of ongoing, progressive shift in neuro-science has allowed one metaphysical *uber-belief* after another to fall by the side.

A few days ago I had a conversation with a bright, articulate woman who proudly proclaimed that she was highly religious and highly addicted to her religious beliefs about her future in heaven. I mentioned that acknowledging an addiction was usually the first step to recovery. "But I don't want to recover. I like the idea of 'pie in the sky' by and by." She said laughingly. I felt enlightened as I realized that it is not the *acknowledgement* of an addiction but the *desire to recover* that is the first step to healing. The catch 22 is that most of the time *addiction,* in and of itself, <u>prohibits</u> the *desiring or wanting* to recover. Anyone ever involved in REHAB knows about that one. (Almost everyone has one in the family.) And besides *Neuro-nastics* is tough!

However, it is a bit easier to understand why religious *uber-believers* have a harder time with belief-addiction-recovery. For example, if a follower of Islam recants his faith he should be put to death according to one interpretation of Islamic law.[2] For some Christians, excommunication or being dis-fellowshipped from their church may seem like a fate worse than death—especially if their friends, social life and recreation have been centered in a church. Typically, the more stringent (fundamental) the religious rules are, the more remorse an adherent feels when turning away—thus there is more fear about making a decision to abandon their addiction. This kind of behavior is predicted by Cognitive Dissonance.[3] But, and it's a really big BUT, fantasizing about the future deprives us of the only reality we possess—the '*here and the now*'.

POSSIBLE SOLUTIONS TO ADDICTION

Attempting to compromise with an *addiction* is like trying to hold on to the *belief* that simple conservation of water (don't wash your car or water your lawn) will stave off a hundred years of progressive drought. Eventually the water reservoirs will dry up—no matter what or how we try conserving the remaining water. While it is true that compromise can mitigate or postpone the ultimate effects, it is never a final solution to the base problem of drought. Most often, the more we postpone confronting the issues, the more deluded we become in *believing* we can control them.

One of the really profound problems with addiction is that most of the basic emotions related to addiction have been around since prehistoric times. When *flight* and *fright* made up most of our emotional repertoire, haste was essential to survival. Today we know that addictions are not just a simple suppression of an action or even a matter of immediate a *free-will* choice, but rather a total brain and cultural interaction. Having said that, the most important thing to understand about addiction is that it is *mind/brain constructed and maintained*. Therefore, three simple steps can accomplish the beginning REHAB of addictions. The first *Neuro-nastic steps are:*

Awareness of the source. Once we have a general understanding of how our brain manufactures our thoughts and habit *nodules*, we have a decided advantage. We can, by practicing continued awareness, facilitate our own brain's buoyant thought (content) loading. If you have an understanding of just how habits are created and how your brain works, you can plot an optimal route through the ever-increasing complexity and '*downers*' of modern life. Consequently, students of the human mind like, Hebb, Holmes, Skinner, and Harris forged new brain paradigms. The most successful brain entrepreneurs were and are those whose mental awareness is kept on affirmative activity and the loading of positive mental content. The more we understand this, the less likely we are to allow unwanted *brain nodules* to control us.

Generate a shift in thinking. Whatever lines of reasoning you have used to explain what you consider habitual, mysterious, extra this or that, paranormal or spooky, suspend it! Delete-delete! Recognize that sometimes before a new insight can be gained, a whole new paradigm shift is required. Our human history is replete with examples of how entrenched ideas and *uber-beliefs* have held back innovation, invention and *brain development*. There is little doubt that just the simple of act of holding on to an *uber-belief* has caused the demise of many of an established civilization let alone single individuals. The *incomprehensible* can become *comprehensible* if you will allow curiosity rather than fixation to dominate your thinking.

Eliminate the reason for a brain nodule in the first place. Often a habit will persist because we refuse to allow our remarkable brain the opportunity to expand and evolve by entertaining new ideas. Even within our current evolutionary limits, we have an almost unlimited source of cognitive ability. We have not begun to discover the boundaries of our amazing brainpower. Even a brief glance at human history is enough to show that we have repeatedly halted human progression by choosing to *believe* rather than *find out*. Today, we have the ideas and technologies to unravel the challenges of this age.

Neuro-nastics, like gym-nastics, means quite literally that repeated exercise changes physical structures. You want to jump on the Pummel Horse and do Olympic-like "flairs". You want to do floor exercises worthy of a Nastia Liukin team member? Well, guess what? That takes a minimum of three thousand hours of training for an already fit athlete. A similar amount of time is required for new brain *nodules* to fully develop. Jump on that *'nastic'* horse and go for it!

AFTERWORD

Summary and the alleged last word

At the beginning of this book there were few rhetorical questions about obtaining "understanding". We were reminded that Einstein once said, "The most *incomprehensible* thing about this world is that it is all *comprehensible*". His implied proviso seems to be "provided we really want to comprehend." Modern neuroscientists and geneticists are rapidly making last year's mysterious unknowable's—considered by half the American citizenry as *uber-beliefs*—into common ordinary c*omprehensible stuff*. Genuine *understanding* is always progressive—i.e. ongoing and unending. Our brain will continue to evolve if we give it half a chance by maintaining our *curiosity*. GS. (*Good Science*)

The most prolific thinkers of any culture in any generation are curious and continuously endeavor to pull their contemporaries forward into *'new' ever-evolving understanding*. The *'Hunter/Gatherers'* did it by migrating to new territories. The *'Farmers of*

Mesopotamia' did it with new crops. The *'Renaissance Men'* did it with new art and ideas. The *'American Founding Fathers'* did it by separating 'king' and country, church and state. *A. Lincoln* did it by challenging entrenched thinking about slavery. *Susan B. Ant*hony led the way to women's rights by changing thought patterns of the male dominated American society. Every intellectual entrepreneur in history, from Hammurabi to Einstein, moved his or her generation forward into new kinds thought and understanding. If Aristotle, S. Freud, T. Edison, R.W. Emerson, E. Holmes, or any of the other *movers and shakers* of their day, were alive today, they would not be merely perpetually repeating their insights in some redundant lecture venue. Creative thinkers and educators help their contemporaries to evolve by making the new thought and new sciences understandable to those who <u>want</u> to *'understand'*.

While it may seem that the *understanding* mentioned in the first chapter is just another illusive *pot of gold*, given *time* and *space* to continue growing, Homo sapiens are figuring the big *"IT"* out. But, there are still many open-ended questions about seemingly ambiguous topics that have not been fully realized. Let me provide a very common one: *pursuing happiness.* TISC ! While Thomas Jefferson said we have the *right* to pursue *happiness*, grasping and/or hanging on is an entirely different matter! Then too, it seems that optimal moments of either *happiness or understanding*, like optimal orgasms or profound epiphanies are few and far between. Furthermore, to be in a constant orgiastic state could be painful. (*"If you have an erection lasting more than four hours, seek your doctor immediately!"*) For sure! TFS.

Contentment, it seems to me, is a kind of balance, or equilibrium. It's far more accessible than happiness. It's knowing that inhale is followed by exhale. It's knowing that when you ingest a beautiful bowl of fruit, you will have a pile of crap; together—nourishment and waste—they represent the concord of life. This kind of balance is important to the very core process of life—inhaling AND exhaling—no dualism here! This kind of *knowing* can be one of life's most elegant epiphanies.

Love represents still another kind of balance or equilibrium. Within the *Golden Rule,* (Love your neighbor as yourself) 'balance' is key. It is not just some sentimental gush—it represents ultimate equilibrium even at the core of lite: The Proton exists for the Electron and via versa. Karen Armstrong's interesting take on *Love your enemies* is illuminating. For example, the word "*love* in Leviticus is a legal text and 'sentiment' is out of place—as it would be in a Supreme Court ruling. *Love* in that context was a legal term used in international treaties. Two kings promised to 'love' each other. This did not mean that they become bosom buddies and exchanged bank accounts but rather that they would be loyal to each other, give each other practical support, come to one another's aid, even if this conflicted with their short-term self-interest." [*K. Armstrong: Compassionate Life. 2010.*] This kind of love results in commitment and *contentment*; it is a kind of balance between self-interest and altruism. Perhaps, this kind of *love* is also the solution for today's global village.

UNDERSTANDING BALANCE

While we may only have a baby's grasp of the full meaning of *love* or life's balance, it is an important beginning, because we are finally beginning to understand that as we accept our authentic tears we can know genuine joy. Some others have suggested that obtaining such balance eliminates our seeming need for traditional dualism and at the same time enhances *curiosity*. Understanding the balance contained in contentment and commitment becomes the process of authentic living. It is the same as inhaling AND exhaling, anabolism AND catabolism, birth AND death—to comprehend this, is the bona fide life gift. It is a painless eradication of dualism. The challenge is how to accept and extol this gift without seeming arrogant or judgmental. Of course, if we can offer some reality based answers along with our questions—that too can be balanced contentment.

MORE QUESTIONS

Perhaps it is good that Santa Claus, anthropomorphic gods, and promises of eternal rewards "IF you do such and such" have been inserted into our human history. They can calm us as children and sometimes control us as adults. But, as the Buddha said, *"After enlightenment—the laundry."* Since we now know that *nodules* (*contend loading of the brain*) control of a whole lot of our ongoing *'think and behavior processes'* we can have *understanding* and also take action to change the physical structure of our brain. WOW! If we can modify human brain *nodules—we can* change human thinking. When we change *thought* we modify *nodules*! But again, we must ask: Is this kind of change really possible with so many bright, articulate people supporting the fixed concepts of ancient mythology as if they were real and <u>benign</u>? We have certainly seen "change" stymied on the national political level! SWEIN.

What usually *wears* us out at any given moment is trying to reconcile the irreconcilable. To resolve this conundrum it is necessary to be able to say *goodbye* before we are able to say *hello*. This is the heart of genuine commitment. Unless one is ABLE to say goodbye, we will always being playing an inauthentic role. Hemingway once said "Optimistic belief keeps a fool from accepting failure." I've been there and done that—without recognizing that failure is the first step to the next success. Being able to say goodbye allows me to say an honest hello.

A friend of mine recently wrote me and asked another kind of question: "Is there a God or isn't there? Since the Big Bang supposition has not as yet been replicated in the laboratory whence did life originate? (Oh, but it has: *The CERN particle accelerator fulfilled that request November 2010.*) I have yet to hear a useful answer. So, until I receive one, I shall be like a stubborn child and believe what I like best, which is—that somewhere, somehow (*our/my*) existence continues. I love my life and earth's beauty too much to settle for *non-existence* throughout eternity. In the meantime, I pray when I am desperate because prayer calms me down and I live in the hope that my health will hold up so that I can continue my extraordinary presence."

History is full to overflowing with examples of just such individuals and their staunch convictions. And because of just such *uber-beliefs*, history is also full of impedes to human cognitive progress. Can we today realize how difficult it was for human beings in the golden age of Greece to fathom the earth as a globe suspended in space rather than held on the back of Atlas? Then there was Heliocentric scientific data provided by Galileo, which flew in the face of *uber-believers*. And later, who would dare acknowledge the simple existence of germs or deny that bloodletting wasn't the most beneficial thing for curing everything. These are only a few of the most obvious obstacles to human cognitive progress.

Once an *uber-belief* takes over, it is ever so difficult to imagine any other alternative. We continue to persist in *believing in* myths and legends even when the related evidence clangs with cacophonous falseness. Such is the power of a *nodule*.

John Stuart Mills once wrote "*anyone who doesn't constantly question will see beliefs turn into worthlessness*" and Bertrand Russell said: "The will to believe is not needed . . . but rather the will to find out, which is the exact opposite." Declaring '*just have faith'* is no longer meaningful in our age of *Google* and Internet research. We don't need anymore dumb-ing down; we need *finding out*! Keeping the *body* and *brain* stretched results in the optimizing of a balanced life. In our *personal stretch* we will of necessity be curious and ask questions and if we listen we will find fulfilling answers. Our brain, the *harbor of consciousness* is also our *spiritual vortex* and as such provokes *curiosity* to whirl and create momentary vacuums, require filling. Fill it we must—with new kinds of wonder and caring actions and the *awe of* fresh new innovative thought!

Do we dare ask why so many are so *comfortable* and find such *peace* in the repetition of the status quo? Today, with the help of neuro-scans, we know that by repetitive action and thought, we morph *nodules* in our brains. They grow to confine or release us. So how we *load the content* of our brain on a daily basis determines how we progress or regress cognitively. In short, we are the progenitors of human brain evolution.

FREEDOM, FREE WILL AND COMFORT

One of the reasons some people in America find the concepts of *comfort* and *eternal life in paradise* so <u>wanting</u> as a reason for keeping the status quo, is that the affluent, educated, 9/11 terrorist pilots who flew into the towers were undoubtedly '*comfortable*' in doing so. They had *faith* that they were going to continue their "extraordinary lives in paradise." When we find such *comfort* in our *uber-beliefs* then it is logical and fair that we allow others to find similar comfort in their beliefs and rituals. In other words, this reasoning supports all sorts of believers, especially those who also yearn for an egocentric future paradise. That's what we all need to object to! We need to avoid supporting non-sense—anyone's nonsense!

How to do this without malice or condescension? A beginning point might be to confront the power of *nodules*—rather than censure the brainwashed human beings we tag with so-called *free will*. If we understand this, we would be less likely to judge others so quickly. Few, if any, rational persons would attribute *free will to the 'weather'* because we know that IF we knew all the variables involved we could predict tomorrow's weather conditions perfectly *i.e. 100% certainty*. But since we also know that to date, meteorologists always miss a few unknowns, the weather is predicted within certain parameters of probability. "There is a 60% chance of rain . . ." etc. Just so, we humans are in fact, unable to account for the unknown variables influencing what we call—human choice. Variables, (genetics, birth trauma, hormones, emotions, environment, toxins, etc.) are constantly impinging and controlling our decisions in a manner that takes a toll. "Free will" only appears seamless—<u>as if</u> we are in control. To become even partially aware of the inexorable links to the limitless chains of cause and effect, which preceded and will follow our brief existence is deeply humbling. And here's the kicker: As Spinoza said, "This teaches us to hate no one, to despise no one, to mock on one, to be angry with no one. And finally, to envy no one." The neighbor who drives you bonkers is—as you may now consider—a function of hundreds of pre-formative and current conditions—yes, including <u>you</u> as a neighbor!

Without awareness and diligence it becomes increasingly difficult with each repeated thought to break free from the stalemate of fixation. As neuroscience has revealed, so-called *free will* is mostly a religious construct necessary to theologically based dogma such as '*sin*'. From another point of view, we probably had no more choice about most of our current *believe systems* than we do about the color of our skin, hair or eyes. As various brain scans testify, indoctrination and habit are powerful in their ability to physically modify our malleable brains. Historically, once fixed brain *nodules* form, we burn the *Bruno's* and imprison the *Galileo's* for suggesting that "*little ol' us*" are not comfortably seated at the center of all that is.

ETHICS, MORALITY AND BRAIN EVOLUTION

Some say that besides comfort, traditional religious principles can provide an *ethical compass* for how to live in this world and even how to treat our neighbor. Why then, do so many religions still perpetuate the oppression of women and non-believers? (Various forms of hell and damnation are still popular venues in both Christian and Muslim worship services.)

Christian scriptural words as per St Paul says "women must be silent", while the Jewish scriptures in Deuteronomy declares that if a woman's hymen is not intact on the wedding night "the men of her town shall stone her to death." Negative sentiments directed at non-believers, as well as women are even more pronounced in the *sacred* Islamic Hadiths. The onus is on religion; human rights should be sacrosanct and not dependent on genitalia or gender. What about women in the highly religious Mid East? More specifically, what about women in East/West negotiations for peace in Iraq and Afghanistan? Will the post war/post revolutionary world still keep the feminine gender as "persons non grata"? Anyone who knows a smidgeon of anything about human history knows that great social changes are impossible without feminine upheaval! Social progress, civility, and genuine democracy can be calculated precisely by the societal position of the so-called fair sex within a particular populace.

Even more to the point, as long as men in power refuse to allow women the right to education, the choice to work in the public sector, the autonomy to walk or drive, in other words the freedom to be fully human, these authoritarian males reveal their own insecurity and failing. This is not just a societal more, it is an amoral "slave mentality" revelatory of male insufficiency and shortfall in every human dimension save physical strength. Yes. In every testable dimension of human aptitude and ability, females are equal, and in most cases, superior to males. No wonder some males are so threatened; we are indeed the inferior gender! Obviously it is no accident that nations that murder women for breaking primitive *male derived* rules are the same ones that create and send "true believers" to progressive countries to slaughter civilians in the name of their god. Truly, there can be no real progress toward world peace as long as men in positions of power and leadership insist on treating women as sub-citizens.

But is this attitude amoral? How do we arrive at what is ethical or moral? Such 'arrivals' appear to vary from generation to generation from culture to culture. 'Right and wrong' appear to contain many interpretations.

Recently, Liane Young an MIT researcher, told BBC. com that scientists have discovered the area of the brain responsible for moral reasoning. Young said that when a simple magnetic field is applied near the temporal/parietal junction of the brain, it changes human ability to make typical moral judgments based upon obvious intention. (e.g. One of the usual distinctions courthouse juries are asked to make is: "Was it an honest mistake or was it intentional?")

Massachusetts Institute of Technology researchers in a attempt to understand how the right temporal/parietal junction (RTPJ) functions asked subjects to evaluate the "morality" of acts depicted in a variety of stories. E.g. A woman gives her best friend what she thinks is a spoonful of sugar in her tea, it turns out the sugar is really poison and the friend dies. While research subjects listened, the MIT techs applied a painless magnetic pulse to the RTPJ area. With this kind of magnetic interference to the area the subjects were inclined to condemn the woman offering the sugar without consideration of her motive.

In another scenario participants were asked how acceptable it was for a man to let his girlfriend walk across a bridge he knew to be unsafe. After receiving a 500 millisecond magnetic pulse to the RTPJ, the volunteers delivered verdicts based on outcome rather than 'moral' principle. That is, if the girlfriend made it across the bridge safely, her boyfriend was not seen as having done anything wrong.

In effect, the magnetic pulse seemed to render them unable to make judgments that required an understanding of *intentions*. In this research, using painless magnetic interference, people's moral judgment appeared to change.

Question: *Would you require a jury of your twelve peers to NOT use cell phones or even be around other types of magnetic interferences before passing judgment on you?*

SELF EVALUATION AND CONSCIOUSNESS

What about this alleged "profundity" that we hominids keep returning to—namely that we are "self reflective" and therefore THE uniquely "conscious" creatures in the universe? But is this really a profound unfathomable concept? OR is the "ME", spoken of in the first chapters of this book, merely an evolutionary defense built in as a necessary part of the ever-enlarging-evolving pre-frontal cortex? In this sense the "I or ME" may be considered as just another genetic/biological construct developed so that we may have an additional "protective" mechanism against abstract future possibilities that the neo-cortex seems fond of revealing. Or are these abstractions just problems that the new cortex invents to keep itself entertained or distracted?

Anxiety, for example, can be considered useful in certain contexts. We use a form of mild anxiety when we take tests, solve problems or predict future possibilities on the basis of mathematical probabilities. After all, we have always known that we are a *pattern seeking* species, which as any paleontologist will tell you, has saved our "species bacon" on many an occasion. (E.g.

Red hourglass on spiders and diamonds on the back of snakes.)

Comfort? Blissful thoughts? At this juncture in our world history it no longer seems appropriate to "wrap drapery and go sustained and soothed" (W.C. Bryant) but perhaps we should rather "rage, rage against the dying of the light." (D. Thomas) When individuals seek the comfort of fixated, non-rational *nodules* they dim the light for all of us. How to brighten the light for others without distain or arrogance is the question.

PLAYING GOD OR A NEW CHALLENGE

We have landed on the planet Mars as well as our Moon. Life has been replicated in the laboratory. And genetic scientists have built the first complete DNA for a one-celled organism from the ground up. These new laboratory creations are especially angst provoking for some *uber-believers* since non-believers created them. In the next few months ever-curious scientists will undoubtedly uncover still more myth challenging facts. E.g. Researchers will soon, painstakingly take chemicals and construct an artificial piece of DNA, insert it into a living cell in order to create new self-replicating entities that is expected to launch an industrial bio-revolution. How real is that prospect? Exxon has $600 million invested in this research. You be the judge.

That scientists would make an artificial living cell and that their work would be would be hailed with jeremiads about Pandora's box, Frankenstein's monster and playing God was predictable, but . . . enough is enough! If you think about it, every technological move forward is a form of playing God, or none of them is—either we want to learn more about the cosmos and employ even greater technologies, or we shut them down with the cheap catchphrase: "He's playing God". Genetic engineering, cloning, stem cell research, fertility treatments and, of course, fetal screening are all called "playing God". Going back in history, people said it about splitting the atom, about organ transplants, about cosmetic surgery. Not that all these technologies and their applications should be free from ethical questioning, but complaining that they are *playing God* is just not helpful. When our distant ancestors first domesticated animals, or grew crops from seeds, or tamed fire, some tribal witch doctor undoubtedly warned them that there were powers that man was not meant to mess with.

Today scientists are creaking open the most profound doors in humanity's history. The creation of viable life forms led Julian Savulescu, ethics professor at Oxford University to say, "This is creating whole new living things from inanimate parts. I think we're going to see a metaphysical earthquake." Biologist Sir Paul Nurse said, "Imagine, say, allying synthetic biology with the genome of Neanderthal man that was described earlier this year. There is much excitement at the idea of comparing this with the DNA of modern humans, in the hope of finding the essential differences between the two—how much more exciting it would be, to create a living Neanderthal and just ask him. And if that seems too morally fraught, may we interest you in a mammoth." TMI **?**

SACREDNESS AND DYNAMIC ANSWERS

Ultimate Answers? Only the fundamentalist *true believers* imagine they have those. The audacious scientists only have brief temporary answers and then more questions. Their research is never finished anymore than the ongoing process of life in the universe is ever finished. Physicists, psychologists and biologists disagree, they argue and debate and they make new discoveries that may contradict previous data. What they do not do is pronounce their data *sacred or the end to end all*, or sanctimoniously proclaim, "just trust . . . take it on faith." A fait accompli is the domain of *uber-believers*.

Good science (GS) always reveals the possibility of something more. If the data shows the earth circles the sun, (and is not the center of the universe set there by some fawning father figure) let's not punish the scientists, but rather say, "OK, what's next?" More questions—bring 'em on! Sometimes it seem we hesitate to ask for fear of literally exploding our comfortable *uber-belief base*. This kind of thinking is BS! *(Bad Science)*.

I do not consider myself to be an atheist in the strictest sense. At the moment I see the developing '*emergence of life*' as *sacred*. I see reflected in the elegant/complex systems of the cosmos another kind of *reverence*. However, these observations do not require anthropomorphic Gods interacting with mankind like a mythological Santa Claus. Thanks to my attorney and friend Don Jordon, I now call myself a *naturalist* as contrasted with a *super-naturalist*.

As noted repeatedly in this book, when we humans repeat any activity or thought process often enough or long enough the physical structures of the brain begin to change. Extrapolating then, this book suggests that when any of us spend a childhood or even a couple of years intensely studying or immersing ourselves in a particular philosophy or discipline, certain aspects of our brain tissue are physically changed and future thought processes are filtered through these physically altered brain parts. *Free will* flies out of the window. A fundamentalist Muslim view of the world simply cannot take in a Western point of view any more than an average fundamentalist from the West can fully appreciate an Eastern view.

Once any of us have repeatedly practiced a particular thought or action (prayer, alcohol, juggling, commerce, gambling, bible study etc) neuro-science explains that structures of our brain are physically modified. It seems little wonder that other points of view become harder to entertain. I think I am usually curious and *fair-minded* and I personally would like to be open to other points of view—or would I? Or better yet—could I?

Finally, to admit that we are composed of natural matter is to ground our selves in the wholeness of the cosmos. It too is just *stuff!* Obviously we coalesce briefly (by comparison) from the same material as the *stars*, and then flow back into wholeness. Such a concept of self can be ennobling, cosmic, ecological—so much more meaningful than the *many mansions of some nether-land* I was taught as a child. Today's science demonstrates that we are the same matter—atoms, electrons and all the rest of the molecular materials—as the rest of the universe. We are of the natural world. Our brain with its trillion connections is the *harbor of consciousness*. There is no longer a need for extraneous metaphysical accountings. Understanding that we are structurally no different from anything in the rest of the cosmos is to allow ourselves to expand into infinity. That too, seems *sacred* to me.

END NOTES

CHAPTER 1

BRAIN NODULES

(1) As reported in the Sept 17, 2010 issue of THE WEEK. The Swedes allowed 18 participants to pig out on fast food for a month while keeping physical activity to a minimum. A control group did not alter their food or exercise routines. After 6 months the "bingers" lost the 10 pounds they gained during the study. But two years later they gained back over half while the control non-binging group did not show any change in weight. The researchers concluded that just one month of pigging out on bad food makes for a lifelong struggle to fight off fat. BBC.com (I wonder what a brain scan before and after would show. (see notes 6 and 7 below.)

(2) All Taxi drivers given brain scans by scientists at University College London had a larger hippocampus compared with other people. This is a part of the brain associated with navigation in birds and animals. The scientists also found part of the hippocampus grew larger as the taxi drivers spent more time in the job. Dr Eleanor Maguire, who led the research team.

"The hippocampus has changed its structure to accommodate their navigating experience. This is very interesting because we now see there can be structural changes in healthy human brains." The posterior hippocampus was also more developed in taxi drivers who had been in the career for 40 years than in those who had been driving for a shorter period. Dr Maguire's research is published in the US scientific journal, Proceedings of the National Academy of Sciences.

(3) D.O. Hebb (1949) proposed that, when a neuron repeatedly takes part in the activation of another neuron, the efficacy of the connections between these neurons is increased. This was an important milestone in the history of the term "plasticity" in reference to the nervous system. (examined in an article in Nature (Reviews Neuroscience, 4, 1013-1019. The legacy of Donald O. Hebb.)

Credit is given to William James for first adopting the terms (like plastic) to denote changes in nervous paths associated with the establishment of habits. Eugenio Tanzi is acknowledged as the first to identify the articulations between neurons, not yet called synapses, as possible sites of neural plasticity. Ernesto Lugaro first linked neural plasticity with synaptic plasticity; and Cajal for being the first to complement Tanzi's hypothesis with his own hypothesis of plasticity as the result of the formation of new connections between cortical neurons. Evidence is furnished that in the first two decades of the twentieth century it was generally accepted that learning was based on a reduced resistance at exercised synapses, and that neural processes become associated by co-activation. Subsequently these theories fell into disrepute when Lashley's ideas about *mass action* and *functional equipotentiality* of the cortex tended to outmode models of the brain based on neural circuitry. However the synaptic plasticity theory of learning was rehabilitated in the late 1940s when Konorski and particularly Hebb argued successfully that there was no better alternative way to think about the modifiability of the brain. In another article see: CNRS, Institut de Neurobiologie Alfred Fessard—FRC2118, Laboratoire de Neurobiologie Cellulaire et oléculaire—UPR9040, 91198 Gif sur Yvette, France) Synaptic plasticity is said to be the cellular mechanism underlying the phenomena of learning and memory. Much of the research on synaptic plasticity is based on the postulate of Hebb (1949) who proposed that, when a neuron repeatedly takes part in the activation of another neuron, the efficacy of the connections between these neurons is increased. Plasticity has been extensively studied, and often demonstrated through the processes of LTP (Long Term Potentiation) and LTD (Long Term Depression), which represent an increase and a decrease of the efficacy of long-term synaptic transmission.

(4) In this book I will often use "*Nodule*" as a synonym for brain plasticity. I recognize that "Nodule" is not a term usually applied to any of the human brain structures but still the word seems well suited to convey the meaning of "physically modified tissue" as intended in this book.

(5) *Brain, Mind and Behavior*, a component of a college equivalent course, includes an instructor's manual and study guide. The neuroscientists who wrote it have dedicated their textbook to the reassuring assumption that all functions of the brain, normal and abnormal, are "ultimately explainable in terms of [its] basic structural components."

(6) The Human Brain does Change structurally: In support Doctors Marianne Frostig and Phyllis Maslow stated 1979, in an article in the *Journal of Learning Disabilities*, "Neuropsychological research has demonstrated that environmental conditions, including education, affect brain structure and functioning." In their book *Brain, Mind, and Behavior* Floyd E. Bloom, a neuro-pharmacologist, and Arlyne Lazerson, a professional writer specializing in psychology, say, "Experience [learning] can cause physical modifications in the brain." Michael Merzenich of the University of San Francisco confirms this. His work on brain plasticity shows that, while areas of the brain are designated for specific purposes, brain cells and cortical maps do change in response to experience (learning).

(7) Recently, German researchers found that juggling increases the brain size. Arne May, neurologist at the University of Regensburg, and colleagues asked 12 people in their early 20s, most of them women, to learn a classic three-ball juggling trick over three months until they could sustain a performance for at least a minute. Another 12 were in a control group who did not juggle. All the volunteers were given a brain scan with

magnetic resonance imaging (fMRI) at the start of the program, and a second after three months. After this, the juggling group was told not to practice their skills at all for three months, and then a third scan was taken of all 24 volunteers.

The scans found that learning to juggle increased, by about three percent, the volume of gray matter in the mid-temporal area and left posterior intra-parietal sulcus, which are parts of the left hemisphere of the brain that process data from visual motion. Students who had not undergone juggling training showed no such change. After the third scan, by which time many recruits had forgotten how to juggle, the increases in gray matter had partly subsided. "Our results contradict the traditionally held view that the anatomical structure of the adult human brain does not alter, except for changes in morphology caused by aging or pathological conditions."

(8) Researchers at University College London studied the brains of 105 people, 80 of whom were bilingual, and found that learning a new language altered gray matter the same way exercise builds muscles. Gaser and Schlaug found gray matter volume differences in motor, auditory, and visual-spatial brain regions when comparing professional musicians with a matched group of amateur musicians and non-musicians. Gray matter (cortex) volume was highest in professional musicians, intermediate in amateur musicians, and lowest in non-musicians.

It seems that, while stimulation causes brain growth on the one hand, the lack of stimulation, on the other hand, causes a *lack of brain growth*. Doctors Bruce D. Perry and Ronnie Pollard, two researchers at Baylor College of Medicine, found that children raised in severely isolated conditions, where they had minimal exposure to language, touch and social interactions, developed brains 20 to 30 percent smaller than normal for their age.

(9) SPECT scan: A Single Photon Emission Computed Tomography (SPECT) scan is a type of nuclear imaging test that shows how blood flows to tissues and organs.

How does a SPECT scan work?

A SPECT scan integrates two technologies to view your body: computed tomography (CT) and a radioactive material (tracer). The tracer is what allows doctors to see how blood flows to tissues and organs.

Before the SPECT scan, you are injected with a chemical that is a radiolabel, meaning it emits gamma rays that can be detected by the scanner. The computer collects the information emitted by the gamma rays and translates them into two-dimensional cross-sections. These cross-sections can be added back together to form a 3D image of your brain.

The radioisotopes typically used in SPECT to label tracers are iodine-123, technetium-99m, xenon-133, thallium-201, and fluorine-18. These radioactive forms of natural elements will pass safely through your body and be detected by the scanner. Various drugs and other chemicals can be labeled with these isotopes.

The type of tracer used depends on what your doctor wants to measure. For example, if your doctor is looking at a tumor, he or she might use radio-labled glucose (FDG) and watch how it is metabolized by the tumor.

The test differs from a PET scan in that the tracer stays in your blood stream rather than being absorbed by surrounding tissues, thereby limiting the images to areas where blood flows. SPECT scans are cheaper and more readily available than higher resolution PET scans.

(10) Learning Disability (LD) researchers often underestimate the wonderful potential and capacity of the human brain. Compare the idea of a unfixable "reading *dis*ability" with Chafetz's view that "the human mind can learn anything." Litvak says, "the human brain has extensive capacities beyond those normally tapped," and Menninger and Dugan say, "the simplest mind today controls dazzling skills, the very same skills that put the universe itself within our grasp".

Their optimism is confirmed by cases on record in which one of the brain hemispheres of a person was removed surgically and the remaining hemisphere was as able to take over the functions of the removed one. Consider the case of 13-year-old

Brandi Binder, who developed such severe epilepsy that surgeons at UCLA had to remove the entire right side of her brain when she was six. Binder lost virtually all the control she had established over muscles on the left side of her body, the side controlled by the right side of her brain. Yet, after years of therapy ranging from leg lifts to math and music drills, Binder became an A student at the Holmes Middle School in Colorado Springs, Colorado. She loves music, math and art—skills usually associated with the right half of the brain. And while Binder's recuperation is not 100 percent—for example, she never regained the use of her left arm—it comes close.

(11) Robert G. Milton *The True Believers* 2005 See Amazon. com

(12) Louis Mencken was the most prominent newspaperman and book reviewer and political commentator of his day. Mencken was a libertarian before the word came into usage. His prose was clear in some cases perhaps too clear. FDR and the new deal were frequent targets for his barbs. He blasted the insatiable appetite of Americans for nonsense and gaudy sham. His life however was definitely not defined by negativity. He was full of joy and his enthusiasm about life was contagious.

CHAPTER 2

POLLYANNA

(1.) Richard Dawkins is well known for his straightforward criticism of creationism and intelligent design. In his 1986 book *The Blind Watchmaker*, he argued against the watchmaker analogy, (an argument for the existence of a supernatural creator based upon the complexity of living organisms.) Instead, he described evolutionary processes as analogous to a *blind* watch-maker.

Dawkins is an atheist, secular humanist. He has been referred to in the media as "Darwin's Rottweiler", by analogy with English biologist T. H. Huxley, who was known as "Darwin's Bulldog" for his advocacy of Charles Darwin's evolutionary ideas. In his 2006 book *The God Delusion*, Dawkins contends that a supernatural creator does not exist and that faith qualifies as delusion. As of January 2010, the English language version had sold more than 2 million copies.

(2.) The old argument of nature and nurture *from a new perspective"*

However, the nature/nurture arguments may soon be over or at least put up for a long rest by the new genetics and neuro-imaging. We are certainly aware of how important genes are in the physical shaping of life forms—including our own. We know that heights, weight and eye color are genetically derived, but sometimes we overlook the importance of genes in shaping and more abstract qualities such as *personality*. In fact since the 2001 implosion of the Genome we now know that genetic influences are so ubiquitous that rather than asking what is inherited, we should probably ask, "What is not?"

You may remember that you inherited twenty-three chromosomes from your Mom and twenty-three from your Dad. Thus, most of your cells contain these forty-six chromosomes which are long slender DNA molecules which if unraveled would be about six feet in length. And you have 46 of these six-foot strands in every one of your body's cells—actually in the cell's nucleus. When those now famous guys, Crick and Watson unraveled the first DNA "double helix" strands way back in 1953, they suggested the molecule was made of two strands of nucleotides, each in a helix, one going up and the other going down, so that matching base pairs interlocked in the middle of the double helix to keep the distance between the chains constant. Watson and Crick showed that each strand of the DNA molecule was a template for the other. During cell division the two strands separate and on each strand a new "other half" is built, just like the one before. This way your DNA can reproduce itself without changing its structure—except for occasional errors, or mutations. DNA's discovery has been called the most important biological work of the last 100 years, and the Genomic field it opened is rapidly becoming the scientific frontier for the next 100 years.

There are literally hundreds of Gene "*recipes*" strung in helix fashion along each of your chromosomes. Interestingly the vast majority of these are used in the fashioning and functioning of your brain. Each *recipe* is a sequence of Nucleotide base letters that informs a cell how to cook up a particular protein. These proteins make up cells—your '*awesome bod*' and brain. However, depending on the information in the genes, the proteins build *cabbages or kings*. A recent estimate suggests we human beings have about 21-31 thousand of these genes scattered along our chromosomes. But they only make up about five percent of the stuff, the rest (95%) has been called "*junk*" some of which represents the skeletons of dead viruses (see chapter 4) and some is vitally important in turning genes on or off thus manifesting a *phenotype*. Which means how you appear—a handsome dog or just a dog. What probably makes home sapiens special among the higher order of mammals is not the thousands of genes (a peanut has 50) but rather when, where, and how the "*junk*" turns the genes on and off.

Just a minor change in one of your sequential chains of *recipe* nucleotide base letters can misinform another of your cells how to cook up a particular protein product—as if your mother used salt instead of sugar. This means that your protein is peculiar, strange or deviant, which typically leads to a devastating phenotype manifestation. We won't mention yours here, but Sickle cell anemia, hemophilia, Alzheimer's and more than two dozen other diseases have been identified as having irregular genetic components. The actual information contained on the long strands of DNA is called your *genotype.* While *phenotype* speaks about your outward appearance—it can have some environmental input as well—but not tattoos. TMI?

So, from the mixing bowl of genes emerge your physical traits, (e.g. skin and eye color) as well as your persona (e.g. memory and novelty seeking). If one of your gene complexes produce a trait advantage (e.g. disease resistance) you will, when compared to the average bear, will produce more disease resistant off spring who will pass it on to their kids. The following generations will be a little better adapted to meet the diseases of our toxic laden twenty first century. Such incremental improvement in a population's gene pool is known as *evolution.*

Faced with the invading hoards of phenotypes that can spring from the mixtures of genes and environment, it would be impossible to know where to look in your genome to find what would be the cause of your particular problem. GN *great news*! We now have at our disposal "genetic imaging" to help figure out your phenotype. We can actually determine the size and shape of your brain's various phenotype structures such as the amygdalate or the hippocampus (see chapter 2) Then we can evaluate your genes and see how they compare. Instead of questioning you, "How do you feel?" and trying to understand your obtuse answers or body language we can now "open the hood and look around with sophisticated instrumentation and make comparisons between blueprints and performance even while the engine is running." Now we can know what the real scoop is inside and outside.

(3.) The **God gene** hypothesis proposes that we inherit a gene set that predisposes us towards experiences that are defined as spiritual, T*he God Gene: How Faith is Hardwired into our Genes* by Dean Hamer is based on a combination of behavioral genetic, neurobiological and psychological studies. The major arguments are: (1) spirituality can be reduced to a number by psychometric measurements; (2) the underlying tendency to spirituality is partially inherited; (3) part of this heritability can be attributed to the gene called Vmat2 (4) this gene acts by altering brain chemical levels; and (5) spirituality arises and continues because spiritual individuals are favored by Darwinian natural selection.

A number of researchers are highly critical of this theory. In his book, Hamer backs away from the title and main hypotheses by saying, "Just because spirituality is partly genetic doesn't mean it is hardwired."

The **Gay gene**. In the 1990s Dean Hamer began studies of the role of genes in human behavior. In 1993 he published a paper suggesting the existence of genes that predispose men (but not women) toward homosexuality and presented evidence that one of these genes was associated with the Xq28 marker on the X chromosome. This finding was replicated in two studies in the United States but not in several others.

(4.) **The Pollyanna brain**. Research repeatedly shows that people are more likely to show preference for *pleasant* than for unpleasant *stimuli* (*Pollyanna* effect; Matlin & Stang, 1978). And by making pleasant associations, (according to an article in Mayo Clinic Proceedings), makes a difference in human longevity. The difference between optimists and pessimists appears to amount to about 12 years of life. The Mayo team began by examining personality tests performed in the early and mid '60s. They proceeded to look well into the future to see how things turned out. Actually they followed their subjects (about 30 years), to measure the relationship between attitude and longevity. Simply stated, they've shown what many have known all along—the mind and body are an inseparable team! Dr. Toshihiko Maruta, the study's chief investigator stated, "The important thing is that we've made a correlation between how people see the world when they're young and how they turn out 30 years later."

(5.) As suggested repeatedly—when we repeat thoughts and/or actions we change the the actual physical structure of our malleable brain. I have used the word '*Nodule*' for this physical change in brain structure. A quick Google resulted in: The credo of this recent revolution is neuro-plasticity—the discovery that the human brain is as malleable as a lump of wet clay not only in infancy but well into hoary old age. In classic neuroscience, the adult brain was considered an immutable machine, as wonderfully precise as a clock in a locked case. Every part had a specific purpose, none could be replaced or repaired, and the machine was destined to tick in unchanging rhythm until its gears corroded with age. Now sophisticated experimental techniques suggest the brain is more like a Disney-esque animated sea creature, constantly oozing in various directions. The Brain is apparently able to respond to injury with striking functional reorganization, and can at times actually think itself into a new anatomic configuration, in a kind of word-made-flesh outcome far more characteristic of Lourdes than a modern scientific theory.

(6.) President Obama utilized "positive thinking" as his election campaign mantra. "Yes we can!" and the word "Hope" were considered essential in his election. "Yes We Can" is also a lyric inspired by our US President popularization of the slogan "Yes we can" during his presidential campaign in 2008.

(7.) Is a person, of average or below average intelligence more easily duped? As I remember, the Bell curve suggests that half of the world's population is below average intelligence. Below and IQ of 100. Imagine that. The "dupe" gene and/or *nodule* is indeed powerful, powerful enough to create mass hysteria and mass belief systems. Many hold that the oldest religious system must be the "true" one, right? At the same moment we all intuitively know that the oldest political system is most likely the most corrupt. Duped again.

(8.) Google a lab rat study done in the 1950's by a scientist that some say was possibly sadistic? (Richter) He took a bunch of rats and put them into a high-sided bucket of circulating water that they couldn't escape from and timed how long it took for the rats to drown. It wasn't long—an average of 15 minutes for the rats to give up, and drown. He then repeated the experiment with a new group of rats and a new twist—in the second instance, he "rescued" the rats just after they had given up swimming, again, at around the 15 minute mark. He let them dry off; he fed them some food, allowed them to recuperate. And then he threw them back in the bucket of water. The amazing result was that these rats were then able to swim for up to 60 hours before giving up and drowning. Richter attributed the rat's newfound stamina and survival skills to "*hope*" and felt that he had demonstrated the miraculous achievements possible as long as one has *hope*.

(9) Voltaire, one of the all-time greatest thought provocateurs said, "Those who can make you believe absurdities can also make you commit atrocities." He knew about the "*duping*" of the masses first hand. Of course he didn't know about the Genome's influence or brain plasticity or *nodules* resulting from repeated indoctrination in the *believing* process. He was certainly a cultivator of diverse ideas. He wanted sundry "idea gardens" to germinate and grow even when others tended them. His humorously satirical writing on religion was in a sense, optimistic, not only in its gaiety but also in his declaration that it was laughable to expect anything but more *duping* to emerge from the popular cleric teachers of the day.

He certainly knew that superstition would cyclically wreak havoc in his "thoughtful gardens". It is fun to listen to him rant about the trendy *dupers* of his day. Especially in our age of Political Correctness and deference to faith, any attack on any "religious belief"—(no matter how absurd produces a gasp and a gap of embarrassing silence or an outraged censor.) Voltaire's glee-filled writings however, may help remind us that an impious spirit and the freethinking that inspired Mark Twain and Henry Mencken needs to be a part of our American literary fabric. (It has almost vanished) At the end of Voltaire's journey, devout Christians dressed his body, as if he were still alive, created a pious parade to *dupe* the populace into thinking that he had recanted. They then moved it to "unsanctified" ground and buried him. Their pretense may have been a joke to them but in the final analysis the joke is on the *dupers*—he lives—his provocative "thought gardens" still flourish. Some of Voltaire's gardening:

"In general the art of government consists in taking as much money as possible from one class of citizens to give it to the other.

The art of medicine consists in amusing the patient while nature cures the disease.

I have never made but one prayer to God, a very short one: "*O Lord, make my enemies ridiculous.*" And God granted it.

No snowflake in an avalanche ever feels responsible.

If God created us in his own image, we have more than reciprocated.

Doubt is not a pleasant condition, but *certainty* is absurd.

Those who can make you believe absurdities can make you commit atrocities.

Common sense is not so common.

If you have two religions in your land, the two will cut each other's throats.

God is a comedian playing to an audience too afraid to laugh.

I may not agree with what you say, but I will defend to the death your right to say it.

It is dangerous to be right when the government is wrong.

A great many laws in a country, like many physicians, is a sign of malady.

England has forty-two religions and only two sauces.

Divorce is probably of nearly the same date as marriage. I believe, however, that marriage is some weeks the more ancient.

Appreciation is a wonderful thing: it makes what is excellent in others belong to us as well.

Anything that is too stupid to be spoken is sung."

Said by Voltaire on his deathbed, to a priest asking that he renounce Satan: "Now, now my good man, this is no time for making enemies."

CHAPTER 3

SILLINESS

(1) Rather than viewing the demise of dominant companies as a function of technological change and entrepreneurial spinning some see a different tale altogether. In the 70s Ames department stores and Wal-Mart were more or less like clones. By 2000 Ames was but a memory while Wal-Mart took one of the top spots on the Fortune 500. Their fates were not a function of technology or market spinning. Wal-Mart's strategy was steady percentage growth. Ames opted for growth via acquisition. The lesson: Thriving or dying depends more on what you do than on what "they" do to you.

(2) Previous to the unraveling of the Human Genome the notion that psychological forces outside human control determine behavior, had eroded our so-called 'moral responsibility' to the point of absurdum-ad-fin-item-ad-nausea. For example, it is alleged (and true) that a defendant once claimed the "Twinkie defense". In that the sugar in the Twinkie made him so hyper that he was compelled to commit murder. Anyone

who's heard of Dan White—and there are fewer and fewer people who have—a clean-cut, conservative San Francisco supervisor who received a light sentence (five years) in the shooting deaths of progressive San Francisco Mayor George Miscode and gay Supervisor Harvey Milk a quarter century ago. For some, the faint memory brings an automatic response: the "Twinkie defense." The impressionable jury, they say, accepted the defense contention that Dan White gobbled Twinkies, which blasted sugar through his arteries and drove him into a murderous frenzy. About as simple as: "Eat a Twinkie, commit a murder." "The devil made me do it" . . . dressed up in psycho-nutritional garb.

(3) News is what it is called. Whatever is "attention grabbing" is the reality. Most of the time what grabs our attention is pain, tragedy or horror. These are only a few . . . for a hundred others check the nightly news.

(4) Ninety-minutes in heaven and on to Twenty-three minutes in Hell. Check Amazon.com for literally hundreds of "inspirational" books to provide comfort related to the experience of death. Hope springs eternal.

(5) See *Steven Pinker, 2002 The Blank Slate: The Modern Denial of Human Nature.* New York: Penguin Books. He is a native of Montreal, received his BA from McGill University in 1976 and his PhD in psychology from Harvard in 1979. After teaching at MIT for 21 years, he returned to Harvard in 2003, Pinker's experimental research on cognition and

language won the Troland Award from the National Academy of Sciences, the Henry Dale Prize from the Royal Institute of Great Britain, and two prizes from the American Psychological Association. His books make complexity seem unproblematic. He is controversial but isn't that better than "PC" conformity?

(6) In the 1930s Ernest Holmes started a whole new trend in the Religious Science movement by taking people where the were (entrenched in traditional types of Christian beliefs) and brought them forward into the new science (the science of a half century ago) An interesting parallel can be made by comparing how kids today are getting their parents to quit smoking.

(7) Twin studies of Homosexuality for instance: Samples of twins recruited from gay and lesbian associations, are possibly biased by the nature of twins who volunteer, but even so, if one identical twin was homosexual, only about half the time was the co-twin concordant (i.e. also homosexual).

Other research based on twins who were recruited for other reasons, and only subsequently asked about their sexual orientation gave a concordance rate between identical twins of less than 50%. There have been two major published studies, one based on the famous Minnesota Registry, the other on the Australian Registry. The larger of the two registry studies is the Australian one, done by Bailey, Martin and others at the University of Queensland. Using the 14,000+ Australian twin collections, they found that if one twin was homosexual, 38%

of the time his identical brother was too. For lesbianism the concordance was 30%. Whether 30% or 50% concordance, all the studies agree it is clearly not 100%. The critical factor is that if one identical twin is homosexual, only <u>sometimes</u> is the co-twin homosexual. There is no argument about this in the scientific community.

(8) See: Jared Diamond's *Collapse* In my opinion *Collapse* is destined to take its place as one of the essential books of our time raising the important question of survival. *Collapse* probes the question of what caused some of the great civilizations of the past to collapse and (perhaps) how we can avoid following in their footsteps. Environmental, ecological damage, rapid population growth and demented silly leadership decisions based on mythology and religiosity were a few of the significant factors in the demise of past societies. Today ominous warning signs similar to those reflected in his book are emerging. *What can we learn from their fates?* 'Okay, okay, I get it' can be said more than once while reading <u>*Collapse.*</u> But something else occurred to me. The redundancy of: poison in the soil (especially salt), cutting down all the trees, disrupting the balance in fisheries and coral reefs via dumping waste into rivers, over-farming, and hunting certain species to extinction all speak to today's world. Diamond repeatedly visits these topics hoping (I think) that the reader will cite them when someone says something like 'The earth will take care of itself . . .' Diamond makes a good case for tying social and economic collapse to today's similar excesses so that the

reader cannot escape the fact that in our modern world any societal collapse effects all of us.

Rebecca Costa's take on the collapse of some ancient civilizations has a slightly different twist. She believes that they encountered a cognitive threshold. I.E. The complexity of their civilizations grew beyond their brainpower. The two elements required for collapse (a.) Gridlock (Similar to our government today) and (b.) "Beliefs" and religious ideology took the place of rationality. (Similar scenarios appear to exist today.)

In 1934 Cambridge anthropologist Dr. J. D. Unwin published *Sex and Culture.* In it he examined 86 cultures spanning 5,000 years with regard to the effects of both sexual restraint and sexual abandon. His perspective was strictly secular, and his findings were not based in moralistic dogma. He found, without exception, that cultures that practiced strict monogamy in marital bonds exhibited what he called creative social energy, and reached the zenith of production. Cultures that had no restraint on sexuality, without exception, deteriorated into mediocrity and chaos. In *The Sexual and Economic Foundations of a New Society,* published posthumously, he summarized: "In human records, there is no instance of a society retaining its energy after a complete new generation has inherited a tradition which does not insist on pre-nuptial and post-nuptial continence . . . The evidence is that in the past a class has risen to a position of political dominance because of its great energy and that at the period of its rising, its sexual regulations have always been strict. It has retained its energy and dominated the society so long as its sexual regulations have demanded both pre-nuptial and post-nuptial continence . . .

(9) Areas of the brain delineated as Brodmann's 9-10-11 are sometimes called the frontal forward (future) thinking part of the brain and it is true for most that when these lobes are severed as in the old "pre-frontal lobotomy" the victim often becomes little more than a "feces producing turnip". Neither anxiety nor ability to anticipate was noted.

(10) See: Andrew Newberg's *Why God wont go away.* This fascinating book of original research has created a whole new approach to religious questions. If "wonder" and "god-connection" vanish when a part of the brain is damaged . . . shouldn't it be obvious where "spirituality" originates? His works allow us to think about brain functions and religion in a bracing new light. It now seems obvious that neurology can be more powerful (and dangerous to established religions) than most of us previously realized in defining who we are and where we have come from.

(11) It is well known that sheep are followers. "To bleat or not to bleat that is the question." It seems that some (sheep?) are enthralled by complex mega-mythological explanations (calling forth Deities or Devils) that are beyond their (bleating) comprehension. The fourteenth century theologian William Occam suggested that numinous explanations should be parsimonious. This idea suggests that explanations should

not be any more complex than absolutely necessary. Hence, "Occam's razor", which disallows the multiplying of entities unnecessarily. For example Bishop Occam would never allow Newton's falling apple to be explained by a tug-of-war between demons and cherubs with the demons pulling the apples down and the cherubs pulling upward. "Ah ha," says Isaac with unrequited gusto, "the demons are much stronger!" To which Occam would have undoubtedly replied, "KISS!" Keep It Simple Stupid!

(12) Oliver Sacks the author of *Musicophilia* is a physician and the author of *Awakenings* that inspired the Oscar nominated film staring Robin Williams. He is a professor of Clinical Neurology at Columbia University.

(13) Penn and Teller (well known magicians in Las Vegas) have an entertaining (read rational and laughable) TV program in which the fraudulent claims of "spiritual" con men and women are exposed.

CHAPTER 4

DESIGN?

(1) BBC news reports December 2005: A court in the US has ruled against the teaching of "intelligent design" alongside Darwin's theory of evolution. A group of parents in the Pennsylvania town of Dover had taken the school board to court for demanding biology classes not teach evolution as fact. The authorities wanted to introduce the idea that Earth's life was too complicated to have evolved on its own. Judge John Jones ruled the school board had violated the constitutional ban on teaching religion in public schools. The 11 parents who brought the case argued that teaching intelligent design (ID) was effectively teaching creationism, which is banned. They complained that ID—which argues life must have been helped to develop by an unseen power—was tantamount to religious education. Consequently . . . "Which god?" became the controversy—thus the necessity of enshrining "The separation of church and state" in the US constitution.

The school board argued they had sought to improve science education by exposing pupils to alternatives to Charles Darwin's theory of evolution. But Judge Jones said he had determined that ID was not science and "cannot uncouple itself from its creationist, and thus religious, antecedents". In a 139-page written ruling, the judge said: "Our conclusion today is that it is unconstitutional to teach ID as an alternative to evolution in a public school science classroom." He accused school board members of disguising their true motives for introducing the ID policy. "We find that the secular purposes claimed by the board amount to a pretext for the board's real purpose, which was to promote religion in the public school classroom," he said. He banned the teaching of ID. It was the first ruling of its kind, and set an important precedent in a country where several states have adopted the teaching of ID, Ironically, he adds, it is a somewhat academic ruling in the Dover area since parents there voted last month to replace the school board members who brought in the policy. (That move provoked US TV evangelist Pat Robertson to warn the town-people that they were invoking the wrath of God.) A lawyer for the parents said the ruling was a "real vindication" for those families who challenged the school board. See: BBC's James Coomerasamy in Washington

(2) See Steven Waldman *the Founding Faith* Random House 2008 NY. The 13 colonies, Waldman says, were indeed Christian, most of them indulging in outright persecution to uphold their "Christian" ideals. But the independent United States "was not established as a 'Christian nation.'" When

George Washington was Revolutionary commander in chief, his own writings typically employed nondenominational language, appealing to *providence* rather than to *Christ*. The First Amendment, which, along with its siblings Second through Ten, was among the first business of Congress under the new Constitution, rejected a national religious establishment. States were allowed to maintain their own religious establishments, and some did so for decades, although James Madison had hoped to dismantle even these. Perhaps the strongest supporters of the separation of church and state in the founding era were the communicants of a new, vigorous church, the Baptists. From 1760 to 1778, there were 56 Baptist preachers in Anglican Virginia jailed. When the Rev. James Ireland continued to preach through the window of his cell, two Anglicans put a bench to the wall, stood on it and urinated in his face. At least 14 jailings of Baptists happened in Madison's home county. "Though much scholarship has gone into assessing which Enlightenment philosophers shaped Madison's mind," Waldman says, "what likely influenced him most was not ideas from Europe but persecutions in Virginia."

(3) Quote from New Scientist (see <u>WEEKLY</u> June 6 2008)

(4) Lander's experiment tested a quirky prediction of evolutionary theory: that a harmful mutation is unlikely to persist if it is serious enough to reduce an individual's odds of leaving descendants by an amount that is greater than the number one divided by the population of that species. The

rule proved true not only for mice and chimps, Lander said. A new and still unpublished analysis of the canine genome has found that dogs, whose numbers have historically been greater than those of apes but smaller than for mice, have an intermediate number of harmful mutations—again, just as evolution predicts. It's not clear why this prediction is called "quirky", except that it's simply mathematical, and perhaps therefore more obscure when written in words. It can be restated as follows. Suppose there are N living members of the species, its "population". Then if a given mutation is harmful enough that it has odds of greater than 1 out of N of reducing an individual's chance of leaving descendants, the mutation is unlikely to persist.

Stated yet more simply, if less precisely, any mutation which is sufficiently harmful probably won't persist. A corollary is that species with large populations tend to have fewer harmful mutations. This is because when N is large, the threshold above which a harmful mutation is unlikely to persist is lower, and so a higher percentage are eliminated over time. Consequently, mice would have fewer harmful mutations than dogs, which would in turn have fewer than apes. And humans would have fewer than chimpanzees . . . Science and Reason October 2005

(5) Candice Pert & Blanche O'Neal: The mind not only talks to the body "it" is also inside the body chemicals that mediated emotion, which are for example, found in almost every cell

in the human body. And even in one celled animals such as amoeba. See: Bill Moyer—*Healing and the Mind*.

The notion that the mind is separate from the physical body goes all the way back to Descartes' "intellectual property war" and the peace treaty he made with the Roman Church. He gave the soul and spirit to the Pope and he got to be left alone to play in the contemporary science sandbox of his day. Cogito ergo sum!

Today we know that mind and matter (flesh) are virtually one item. The mind is not just in the brain but rather the mind is part of a communication network throughout the brain and body. With this new information we begin to see how physiology can affect mental functioning as well as visa versa.

While the brain itself is a kind of window to the outside via the eyes, ears, nose and mouth, while the mind is an enlivening information realm throughout the brain and body that allows the cells to talk to each other and the outside to talk to the whole organism. The link or connection between the mind and the body is what we refer to as the "Neuro-peptide message of emotion."

(6) The Roman Catholic Church prohibited the advocacy of heliocentric ideology as factual, because that theory was contrary to the literal meaning of Biblical Scripture. Galileo was eventually forced to recant his heliocentric and spent the last years of his life under arrest on orders of the religiously

based Inquisition. Not only the Roman Catholic prosecuted "heretics"—similar institutions existed within the Calvinist and other Protestant churches.

(7) When bio-scientists in South Korea announced that they had allegedly cloned human embryos in a lab, Rick Weiss (Science Writer, THE WASHINGTON POST) said the report does offer the most detailed recipe yet for making a cloned human embryo—that is an embryo that is made just from a single adult cell. And if you were to take one of these embryos and transfer it to a woman's womb, then, in theory at least, it could develop into a cloned human baby. There is one interesting glitch so far also: in this set of experiments, at least, it only worked with female cells. So with men there is nothing on the horizon through this technique. It set off, or reinvigorated, the debate about what should be done in this country about doing this kind of research—the debate that begins with a moral argument about what these human embryos are. There are many people who believe that research should not be done on human embryos—that they are entities with "moral standing". And, in fact, in this country, where federally funded scientists do the majority of basic biomedical research, this entire workforce is precluded from doing this kind of research.

(8) Daniel Dennett of Tuffs University is one of our most provocative modern philosophers. His Books will provide the reader both wit and understanding. See *Darwin's dangerous Idea.* 1996

(9) The Gustave-Roussy Institute in Paris is part of the French Institute of Health and Medical Research. Best know for cancer research rather than controversy.

(10) Source: <u>Rutgers, the State University of New Jersey</u>, Richard H. Ebright—Professor Ph.D. 1987, Harvard . . . American Society for Biochemistry and Molecular Biology

(11) In arguments over moral issues such as cloning and same-sex unions, those who claim they know best often cite what is believed to be "Natural." But when this premise is examined more closely we see that the diversity of reproductive strategies seen in animals reveals that "there are more things in heaven and earth Horatio than are dreamed of in your [moral] philosophies." Some natural creatures clone themselves, some engage in homosexuality, and still others seem to switch genders. The Komodo dragon can create progeny without benefit of daddy dragons. These "virgin births" observed in zoos are a form of cloning. Several species of fish can also reproduce by cloning, which is apparently an ancient tactic used by a variety of species to live on in a precarious world. In nature some are born male and become female and vise versa. In the natural animal world there is no abnormal—just survival and adaptation. So as we conduction our prolonged and often painful debates over moral and ethical issues—let's leave nature out of it. (See: Neil Shubin The New York Times)

(12) According to Eric Lander, "People today are now living through the most stunning information revolution, unlike anything before in the history of science." He compared its importance to the chemist Mendeleev's critical observation around 1869 that all the elements of matter could be organized in a very simple table. With this discovery, Mendeleev laid the foundations for the chemical industry and for much of chemistry in the 20th century. The biological sciences and industry are now experiencing the same thing. Instead of a periodic table, the human genes constitute a finite list that will be complete in the near future. This list will help biologists and scientists understand the tremendous diversity of the human race and determine the causes of inheritable diseases.

Every possible DNA sequence and DNA change that can exist probably does exist somewhere in the world. On the other hand, there are only two or three common variants of most human genes. If two people were selected at random from any audience and a particular gene were sequenced from each, the odds are one in two or one in three that the two sequences of the coding regions would be identical. This reflects the fact that the human race descended from a small population in Africa only 10,000 generations ago or about 200,000 years. Small populations have relatively few variants, and the mutation rate of one in a billion bases is so low that 95% of all the genes in the audience have not undergone a single mutation in all those years. Even though any two human chromosomes are nearly identical, the little differences in DNA sequence can be used to trace the inheritance pattern of chromosomes and localize particular genes to particular sub-regions. Genetic variations influence our lives, but they don't constrain us, nor do they shape us in the choices we can make as a society. What has happened so far in the information revolution will seem like nothing when compared to what will flow from the sluice gates of human genetics projects around the world over the next decade or so. We must explore "how to manage the information, and the choices and consequences of what science has to offer." Human Genome Program, U.S. Department of Energy, Human Genome News (v10n1-2).

(13) For follow up reading see Robert Milton 2005 *The True Believers*—2005 pp 31-68 & Newberg 2001 *Why God won't go away.*

CHAPTER 5

NEW?

(1) Europe has the historic opportunity to be a positive and mobilizing model for all those countries that are considering regional integration. (e.g. Latin America, Asia and Africa) Europe has the prospect to set in motion an international policy in favor of peace. Today's Europe is capable of opening new horizons and the new paths that humanity needs to take. Europe has the chance to put itself at the head of an historic change, in the vanguard of the Universal Human Nation. This suggests that the answer to world peace is trade and the commercialization of the globe. (i.e. globalization may not be as evil as some think)

(2) Jostein Gaarter is the author of the international bestseller *Sophie's World* that, while it is a whimsical novel, is also a scintillating history of philosophy.

(3) In its most basic form, the thought process in a valid scientific method consists of the following tasks:

Identifying a Problem—Forming a Hypothesis—Designing and Performing Experiments—Collecting and Analyzing Data—Formulating Conclusions about the Hypothesis—repeat (replicate) the experiments and achieve the same observable results. (Optimistically).

In 1907 a Massachusetts physician named Duncan MacDougall tried to find out the weight of a SOUL by weighing six dying patients before and after their death. He reported in the medical journal *American Medicine* that there was a 21-gram difference. Even though his measurements were varying, and no one has been able to replicate his findings, it has nonetheless grown to urban legendary status as the weight of the soul. Thus beliefs come into existence and "stick" in the memories of some gifted, charismatic intellectuals who rarely go back to check the facts but repeat the antidotal stories in a most convincing manner.

(4) Quantum Consciousness. The study of the actions of subatomic particles through quantum mechanics produces what Einstein once called "spooky action at a distance." When the observation of a particle in one location instantaneously affects a related particle at another location (which, while it has been approximated in laboratory conditions, could theoretically mean in another galaxy). Again while in apparent violation of Einstein's upper limit of the speed of light, it is theoretically possible. Theoretically is the key word. Most *New Agers* take this to mean that the universe is one giant quantum field in which everything (and everyone) is interconnected and can

influence one another directly and instantly. They also apply Quantum Mechanics to the study of consciousness to explain how the human brain represents the entire tangible world through biochemical signals. Einstein later acknowledged that Quantum Physics was not spooky, but still later, Physicists agreed that for a system to be defined as Quantum, the mass m, speed v, and distance d (mvd) must be on the order of Planck's constant h. If it (mvd) is greater than h it cannot be treated as a quantum system. The neural system of humans is about three orders of magnitude too large for quantum effect to be influential. See Victor Stenger, Univ. of Colorado. It would seem that Q-M is not applicable. No—not a mix of apples and oranges but more like elephants and microbes.

(5) Robert Merton was one of the most influential sociologists of the 20th century, whose coinage of terms like "self-fulfilling prophecy" and "role models" filtered from his academic pursuits into everyday language. He was 92 when he died in 2003.

Merton gained his pioneering reputation as a sociologist of science, exploring how scientists behave and what it is that motivates, rewards, and intimidates them. By laying out his "ethos of science" in 1942, he replaced the entrenched stereotypical views that had long held scientists to be eccentric geniuses largely unbound by rules or norms. It was this body of work that contributed to Mr. Merton's becoming the first sociologist to win a National Medal of Science in 1994. But his explorations over 70-odd years extended across an extraordinary range of interests that included the workings of the mass media, the anatomy of racism, the social perspectives of "insiders" vs. "outsiders," history, literature and etymology.

Merton's inquiries often bore important consequences in real life as well as in academics. (See Brown vs. Board of Education) He spent much of his professional life at Columbia University, where along with his collaborator of 35 years, Paul F. Lazarsfeld, who died in 1976, he developed the Bureau of Applied Social Research, where the early focus groups originated. The course of his career paralleled the growth and acceptance of sociology as a bona fide academic discipline.

(6) In the 1870s, two inventors Elisha Gray and Alexander Graham Bell both independently invented devices that could transmit speech electrically (the telephone). Both men rushed their respective inventions to the patent office within hours of each other. It seems Bell filed his patent before his Quaker competitor from Illinois. However Elisha Gray and Alexander Graham Bell engaged in a notorious legal battle over the telephone invention—obviously Bell won.

(7) It has been suggested—in oracle-like fashion—that by creating a link between consciousness and quantum Physics "we may find out how the brain might create subtler worlds, the kind traditionally known as heaven. If the secret lies not in brain chemistry but in awareness itself, the afterlife may turn out to be an extension of our present life, not a faraway mystical

world." When questioned the people I asked, (who professed to know such things), had no Idea what this meant.

(8) The Oracle of Delphi (1400 BC) historically the most important shrine in Greece. Built around a *sacred* water spring, Delphi was considered to be the *omphalos*—the center (literally navel) of the world. People came from all over Greece and beyond to have their questions answered. The answers, usually cryptic, and ambiguous could determine the course of everything from when a farmer planted his seedlings, to when an empire declared war. Arguments over interpretations of an oracle were common, but the oracle was always happy to give another prophecy if more gold was provided. A good example: The famous words "a great army shall be defeated today." No mention of which army was given so it could mean either or both. While ambiguity and vagueness are still favorite ploys of fortunetellers today they have become more sophisticated in providing believable readings. Cold reading: Ask a multitude of questions and make numerous statement and see what gives: e.g. I'm getting a T name. Who is this please? I see red. What is this? On and on. Most are wrong but subjects only need an occasional reinforcement to be convinced. Las Vegas depends on this human memory foible for its survival. Warm reading: using list of high probability guesses. I.e. Scar on knee, photo album incomplete, electronic gadget or clock that no longer works, childhood accident involving water, etc. It is remarkably easy to convince. It seems that human beings long to "create great *uber-beliefs*."

(9) Shamanism is traditionally defined as: 1. An animistic religion believed by certain peoples of northern Asia. They believe that mediation between the visible world and the spirit world is real—accomplished by their religious leaders called shamans. 2. A similar religion or set of beliefs, especially among certain Native American peoples.

(10) Kundera, Milan: Born in Czechoslovakia in 1927. He has written nearly a dozen books and in the process raised the ideas of the novel to a new dreamlike lyricism and emotional intensity. See: The *Unbearable Lightness of Being*.

(11) Hatfield and Cacioppo wrote a fascinating treatise on the subject of *Emotional Contagion.* Mimicry subtle and almost invisible is one of the primary ways we Homo Sapiens infect each other with feelings and attitudes. See also the term *Meme*.

(12) Dr. Friedman's research on nonverbal communication at the university of California at Riverside focuses on expressive style and its relation the transmission of emotion. Facial expressions and body movements are employed to examine social influence, and personal charisma. His Affective Communication Test (ACT) available from him for businesses or professionals. Other studies reveal that when people are exposed to emotional facial expressions, they spontaneously react with distinct facial electro-myo-graphic (EMG) reactions in emotion-relevant facial muscles. These reactions reflect, in part, a tendency to mimic the facial stimuli.

Whether corresponding facial reactions can be elicited when people are unconsciously exposed to happy and angry facial expressions is another question. Through use of the backward-masking technique, subjects were prevented from consciously perceiving 30-ms exposures of happy, neutral, and angry target faces, which were immediately followed and masked by neutral faces. Despite the fact that exposure to happy and angry faces was tachistoscopically presented (subliminal), the subjects reacted with distinct facial muscle reactions that corresponded to the happy and angry stimulus faces. Results show that both positive and negative emotional reactions can be 'unconsciously' evoked, and particularly that important aspects of emotional face-to-face communication can occur on subliminal level.

(13) Your body's locked precisely with your speech. You can't break out of this no matter what you do. Your eyes even blink in synchrony with your speech." Movements appear to begin, change, or end on the same film frame that a new vowel or consonant begins—within about four-hundredths of a second in the new sound is translated into muscle movements and then into airways that hit your ear, and your eardrum starts to oscillate in absolute synchrony with the voice. In essence there's no vacuum between—it takes only a few milliseconds for a sound to register in the brain stem, 14 milliseconds for it to reach the left hemisphere. Also see "Cultural Micro-rhythms" in M. Davis (ed.) *Interaction Rhythms: Periodicity in Communicative Behavior.* New York: Human Sciences Press 1982.

(14) The 12 steps to nodule recovery repeated practice for at least 30 days

1. **Admit** that when one person has a fanciful delusion it is called psychosis when a dozen people have the same delusion it is called fundamental religion.
2. **Come** to understand that neither Santa Claus nor any of his helpers are real. They were made up in order to give small children temporary comfort. (And perhaps to control them)
3. **Make** a decision to take charge of your own life and stick by it. If you don't pay your credit card charges they will take it away from you. It is that simple.
4. **Make** a searching and fearless inventory of yourself, recognizing that every human being has a concept of right & wrong—black and white—but gray is sometimes OK too.
5. **Admit it**, openly and unashamedly, when you are wrong. Everybody is at one time or another. But where apologies are absolutely necessary they usually aren't possible.
6. **Decide** what changes you want to make and recognize that sometimes change takes a little time. Losing weight is much better over a year because then it becomes a life style. TTTT
7. **False humility** is a pain in the ass. Most of the time when apologies are possible they are not necessary. "They" love you in spite of yourself not because of yourself.

8. **Go** out of your way to make amends for uncouth behavior. Pay your debts or declare bankruptcy. Remember that good people can do good things and bad people can do bad things but for good people to do bad things it takes religion. See 911
9. **Call** a spade a spade. Be honest, if you have friends who still believe the earth is the center of the universe—enlighten them but don't argue. Some people "need" their illusions.
10. **Maturity** and the enlightenment is the work of a lifetime. Stay at it step-by-step—day-by-day-hour by hour-minute by minute-second by second-
11. **Utilize** a daily routine of meditation and exercise. There are many ways to relieve stress—one of the best is make a decision quickly knowing you can remake it in a heartbeat.
12. **Having** had an epiphany as the result of the aforementioned steps, share these steps—but for crimany sakes don't become an obnoxious pain in the butt rant and roller like me

CHAPTER 6

RELIGIOUS SCIENCE & SCIENCE SCIENCE

(1) Neuro scientists such as Damasio and Newberg are gradually blurring the borders between science and religion or God and Naturalism. *Naturalism* asserts that the world is of a single piece; everything we are and do is included in the space-time continuum whose most basic elements are those described by physics. We are the evolved products of natural selection, which operates without intention, foresight or purpose. Nothing about us escapes being included in the physical universe, or escapes being shaped by the various processes—physical, biological, psychological, and social—that science describes. In the context of the scientific understanding of us, there's no evidence for immaterial souls, spirits, mental essences, or disembodied selves that stand apart from the physical world. See: naturalism.org

(2) Much of what passes for science, especially in religious circles, resolves itself with the help of the mysterious "dupe"

gene. Nostrums are what some believers present as a secret composition highly recommended by its preparer but usually without empirical evidence of its efficacy.

(3) The "Big" Bang Model is broadly accepted as the origin and evolution of our universe. It postulates that 13.7 billion years ago, the portion of the universe we can see was only a few millimeters across. It has since expanded—oooommmmm ad infinitum—from this hot dense state into the vast and much cooler cosmos we currently inhabit. We can see remnants of this hot dense matter as the now very cold cosmic microwave background radiation of the cosmos. Further, the remnant heat (radiation) from the Big Bang has a temperature, which is highly uniform over the entire sky. This fact strongly supports the notion that the gas, which emitted this radiation long ago, was very uniformly distributed.

(4) On the average, a one-year old human has about 100 billion (10^{11}) neurons. Those neurons are lost at a rate of roughly 200,000 per day (a net loss of 2 to 5% by age 50). Alcohol may facilitate this loss. Maximum brain weight is achieved at about age 21—two and a half to three pounds. A typical brain neuron will have connections with at least 1,000 other neurons. Aside from neurons, the other important brain cells are the glial which are a kind of glue and more numerous than neurons (a human brain may contain a trillion + glial cells). The synaptic connections may ultimately prove to be some googol number larger than a trillion trillion.

(5) The miracle of the human body calls for celebration: The physical body is the mediator of all our experience. If it seriously malfunctions, everything we perceive is distorted. If the physical body is relaxed and stress-free, most of our difficulties appear to fade to insignificance. The impact of the body on our awareness is stronger than the influences of history, culture, or psychology; although they, of course, affect it. Individuals vary widely in the amount of attention that they devote to the body, but except in sleep, unconsciousness and death, human beings cannot escape its impact. Its importance is like water to fish easy to overlook but obvious when it is disturbed. John Mann . . . *Body of Light* ch 1 paragraph 1.

(6) See Neuro-ethics: Mapping the Field Conference proceedings—San Francisco 2002, Sponsored by the Dana Foundation. Hosted by Stanford University and University of California, San Francisco. Steven J. Marcus, Editor. THE DANA PRESS, New York NY.

(7) Like religion, our legal system has yet to catch up with neuro science. If brain science can provide a better diagnostic evaluation and better predictive efficacy, these scientific outcomes will eventually and significantly change the way our criminal justice system and its jury of 12 peers provide final judgments—just as DNA science has already accomplished.

(8) Harris, Sam, *The Moral Landscape* (page 221 ff) 2010 NY. Norton Press

P.S. Let's look at a more Buddhist-like conception of morality: *God is the Cosmos and all that is therein. Every star, every blade of grass, every grain of sand—and every person and every thing is a part of the God process including morality. Furthermore, She-He-It is not a single Personage, but All that there is, and I am within She/He/It and She/He/It is within me because my astonishing ever-evolving neuro-system creates and projects <u>all</u> of it* . . . including so called morality and the *Shit* called—She/He/It. LOL?

CHAPTER 7

MODERATION

(1) President Obama's Egypt speech was not particularly well received in either Europe or the US. Commentators were mostly critical in their appraisal before and after:

Mitt Romney—(U.S News) commented: [W]e certainly should not stand up and apologize for America. America has sacrificed too much to restore liberty to people in the world to ever be in a position of constant apology. I think the president was wrong in going on Arabic TV and saying that America has in the past dictated to other nations. I think he was wrong in fact and that it was the wrong thing to say . . . I hope as [Obama] goes to Cairo he shows the resolve and strength of America on preserving and defending freedom and does not in any way suggest an apology. This is a time for strength and commitment to common principles, not a time for apologizing for America. We have done too much. Too many lives have been sacrificed on behalf of the freedom of other people in the world for America to engage in an apology tour. (However, others counter with "It seems that America often preserves freedom in countries where <u>oil</u> or other <u>natural resources</u> are available for exploitation.")

Still others say: Apology is "something he definitely ought to leave at home: grating apologies for America's past . . . Genuine pride in representing America will do just fine. Our nation has no peer in liberating people from the grip of tyranny, especially in the regions Obama will visit. That's a fact of history and the President of the United States ought to take every opportunity to say so . . . He can say loudly and clearly that President Bush conceded it was a mistake to use the word "crusade" to describe the war on terror immediately after 9/11. *New York Daily News* (Perhaps if we bothered to ask why "crusade" would be considered abhorrent in the Mid East countries, we might better understand why an apology could be appropriate.)

(2) The proposed Islamic Community center will have a basketball court and a prayer room but is not technically a Mosque. Furthermore, it is <u>not</u> at ground zero but several block away at the old Burlington coat factory. Most Americans agree that we cannot simultaneously acknowledge a right and at the same time insist that Government abandon it.

But Sarah Palin, and Senate Majority Leader Harry Reid have both offered statements designed to give the appearance of being "tolerant" while adhering to good old-fashioned PC values: *"They have a right but should they?"* Sarah Palin.

". . . *think the mosque should be built some place else*" Harry Reid

There is NO way that individuals have a protected right to do something and simultaneously criticize their own government for supporting the execution of that right.

(3) Political correctness has a basic flaw. If all views are equal, why do some who embrace this view feel the need to push this agenda as the "correct" one at the same time demonizing other views as "incorrect"? This is a satirical example of PC correctness??? "*Jamal was offended by me calling him a perverted gay fairy black boy all strung out on crack. The politically correct thing to say would be that Jamal is leading an acceptable alternative lifestyle as an African-American homosexual who has a drug addiction disease.*" (Google)

(4) *THE WEEK* Oct 10, 2010, p 16 reported that Islamic fundamentalists threaten to bomb any newspaper that offends them. Apparently this same group in America has routinely offered such threats. Evidently, publishers are terrified (oops, there is the word Terror) of publishing anything that might offend Islamic exquisitely tender sensibilities.

(5) Recently there have been many negative comments regarding Islamic education. In most of the Muslim world only boys were(are) formally educated and even then it was(is) primarily a matter of memorizing the Quran. The following is a typical evaluation: "Though the noble Qur'an professes the

importance of education to the entire humanity, the irony is that beneath the lamp it is always dark. On actual count, due to some unfortunate course of action, Muslims are among the most educationally backward communities in India. Except a few states, the performance of Muslims reckon below the national average almost everywhere. While there are many reasons for the community's failure in achieving a good educational profile and maintaining the progress, at least one apparent reason is that the community has, in the first place, failed to take appropriate and adequate initiatives towards education . . . Glibly put, one can say that the community has not cared enough for formal education, rather it has channelized its resources towards religious education by setting up Madrasahs." S.R. Talukdar *A Window into Indian Muslim Life July 10 2010*

(6). The reality is that un-reinforced behavior eventually extinguishes. Las Vegas syndrome: When reward is randomized the behavior will continue much longer without reward or reinforcement. What Radical Christian preachers and Islam Imams have done is to randomly reward each new generation of believers.

(7) In a speech to the UN General Assembly, M. Ahmadinejads, The Iranian president said there was a theory that "some segments within the US government orchestrated the attack to reverse the declining American economy and its grips on the Middle East, in order also to save Israel (he called them

Zionists). To an astonished chamber, he also claimed that the majority of the American people as well as "other nations and politicians agree with this view,". UN General Assembly September 2010.

(8) Google has become the premier search engine on the World Wide Web. Less and less it is necessary to post references and suggested reading at the end of published books, since those who wish to find references or further readings need only to ask Google.

For example, the term "underdeveloped" is defined as "having a low level of economic productivity and technological sophistication". These terms are in turn very subjective. "Economic productivity" is generally measured by GDP, which is actually a measure of the total price charged for all economic activity plus a measure of the aggregate number of hours worked for paid wages. "Technological sophistication" is a measure of how complex (and costly) technologies are, not a measure of their efficacy or value.

Despite this dubious definition and distinction between "developed" and "undeveloped", we in the "developed" West almost blindly accept these arrogant propositions:

1. The "underdeveloped" nations have always been full of misery, suffering, deprivation and abject poverty. "This is their own fault."

2. Their salvation lies in learning the lessons (political, economic, educational, cultural, technological and social) of the West and changing to be just like them.

It is almost ironic that this arrogance prevails at the same time the West is realizing that we are in fact "overdeveloped"—too much waste and pollution, and overuse of soil, water, land, and oil, among other things.

Some of the "developed" world (notably the Americas)—actually began as exploited "underdeveloped" countries. The vast majority of the First Nations of the Americas were killed by genocidal campaigns conducted by Europeans. After awhile the European settlers who stole their land got tired of their colonial status and used their isolation, familiarity with the new land and newer European technology to liberate themselves from their European exploiters, and become exploiters of other countries in their own right.

But the answer to the plight of the "underdeveloped" nations *is not more development*. It is an end to exploitation, tyranny and overpopulation. The point of this is that we need some new terminology: The terms "underdeveloped" and "developed" no longer make sense (if they ever did). The term "Third World" begs the question of why it is still called that when there is no longer a "Second World". I'm tempted to be mischievous and suggest we call the affluent nations the "over-consuming" world and the rest the "overexploited" world (exploited by those both inside and outside the counties, and by exploding populations ditto). But somehow I don't think I could get conservatives to use these terms. We could settle on "rich" and "poor", except that many countries have an abundance of riches but none or just a few of its residents share in that wealth.

(9) Sensory deprivation eventually causes the brain to release endorphins, and then some report a lowering of stress with a sense of well being that often borders on euphoria. Some call it "meeting with God or the Devil." Since John Lilly developed a deprivation tank in 1954 to study the brains reaction to the loss of input, reactions greatly depended on the length of time in isolation and previous experiences in isolation.

(10) Neuro research by Newberg using the SPECT scan was one of the first to demonstrate that "oneness with God" was a function of brain activity. Theological "belief", denomination or creed per se appeared to be of no consequence. See: _Why God Won't Go Away_ Ballantine Books, 2001 New York.

CHAPTER 8

INSPIRED

(1.) *Google* has over 1000 references to "inspired writing". It seems that everyone either wants or already has "inspiration". Once the term is applied to a particular script it takes on a life of it own—especially if the writer is charismatic, a good speaker and can get on the "motivational circuit", which is a *cash cow* for allegedly "inspired" individuals.

(2.) The separation clause is a legal and political principle derived from various documents of several of the Founding Fathers of the United States. "Congress shall make no law respecting an establishment of religion, or prohibiting the free exercise thereof is the key to protect both religion and government . . ." John Locke is often credited for the modern concept but the phrase "separation of church and state" is generally traced to Jefferson's letter to the Baptist church at Danbury. His purpose in this letter was to assuage the fears of the Connecticut Baptists, and so he told them that a wall had been erected between church and state to protect them.

The metaphor was intended, as The U.S. Supreme Court has currently interpreted it since 1947, to mean that religion and government must stay separate for the benefit of both, including the idea that the government must not impose religion on Americans nor create any law requiring it.

(3.) Adherents of the Judeo-Christian Bible and the Muslim Quran claim their books to be divinely inspired and are often referred to as the "word of God/Allah." Both books have resulted in followers willing to die for a particular interpretation. Further, each book has many splinter groups, which make the "God inspired" interpretation questionable. That is, if the books are from God to his creations shouldn't they all be on the same page? Jewish Rabbis debate their differences like there is no tomorrow.

(4.) Karen Armstrong has written more than a dozen scholarly books and has become one of "religion's'" most beloved advocates. And although she draws from a myriad of religious traditions, one learns from her own varied experiences that one grows from adversity (e.g. a failed PhD thesis; years of undiagnosed epilepsy; and, of course, her much-noted years as a Catholic nun). Seemingly Armstrong has adopted a theology of "ortho-praxy", which draws her into compassionate living, honoring and accepting both weaknesses and strengths. Armstrong contends, "<u>Compassion is a habit of mind that is transforming</u>. (creating *nodules?)* One must be prepared to extend compassionate interests where there is no hope of a return." She concludes, based on her exhaustive research of Christianity, Islam and Judaism, that "fundamentalist movements distort the tradition they are trying to defend by emphasizing the belligerent elements in their tradition (alleged inspired scriptures?) and overlooking the insistent and crucial demand for compassion. Armstrong's first memoir, _Through the Narrow Gate_, ended not long after she acted upon her decision to leave the convent where, after seven years "she had become a skeptical nun." _The Spiral Staircase_ picks up her intellectual and religious questing and brings her devoted readers up to date on the result of her explorations into the nature of God and his/her(?) place in our world and lives. The only relationship that seems to matter for Karen is her relationship with God, a being who, in her view, <u>probably does not exist</u>.

(5.) _The True Believers_ Dr. Robert Milton 2005 see Amazon.com

(6.) When the BBC decided in January 2005, to broadcast Jerry Springer, they received over 63,000 complaints by offended Christian viewers who objected to the show's portrayal of Christian icons (including one scene depicting Jesus professing to be "a bit gay"). A fundamentalist group sought a private blasphemy prosecution against the BBC, but the charges were rejected. The Christian Voice applied to have this ruling overturned.

In January 2008, a spokesman for the Prime Minister announced that the Government would consider the abolition of the blasphemy laws during the passage of the Criminal Justice and Immigration Bill. The Government consulted with churches before reaching a decision. In March 2008 Peers voted for the laws to be abandoned.

Relatives of a Pakistani Christian woman sentenced to death for insulting the Prophet Muhammad say they will appeal against her conviction. Asia Bibi is believed to be the first woman sentenced to death under Pakistan's blasphemy law. The 45-year-old mother was sentenced to death by a court in the town of Nankana, around 75km (45 miles) from the city of Lahore in Punjab province.

Her husband told the BBC her conviction was based on "false accusations." Although no one has ever been executed under the law, 10 *accused* have been murdered before the completion of their trials.

CHAPTER 9

AMERICANIZATION

(1) Former *ECONOMIST* editor Bill Emmott said newspapers had played their part in his magazine's success story by preferring 'entertainment' journalism to serious analysis. Mr. Emmott, who has seen the ECONOMIST circulation double to more than 1m during his 13-year editorship, attributed that performance to rigorous analytical skills and a growing appetite for global news from a highly educated world audience. "People want concise authoritative analysis and that's what we offer," Mr. Emmott said today, after announcing his resignation from the post of editor in chief. "Among our big niche of readers, their interest in and exposure to global affairs has increased. More people have business exposure worldwide and more people are affected by global events, whether that's jobs, terrorism or bird flu. And people are better informed than ever—Today there's more of a market for high-quality, well-written analysis."

(2) Khanna in *the Second World*—Parag Khanna is an expert on geopolitics, global governance, and Asian and European affairs, and was most recently the Global Governance Fellow at The Brookings Institution. He has worked at the World Economic Forum in Geneva, Switzerland, where he specialized in scenario and risk planning, and at the Council on Foreign Relations, where he conducted research on terrorism and conflict resolution. He is the author of *The Second World: Empires and Influence in the New Global Order* (Random House, 2008). Mr. Khanna holds bachelors and masters degrees from the School of Foreign Service at Georgetown University, and is completing his PhD at the London School of Economics and Political Science. He speaks German, Hindi, French, Spanish, and basic Arabic. His writings have appeared in *The New York Times*, *The Financial Times*, *Harper's Magazine*, *Policy Review*, *Foreign Policy*, *Prospect* (U.K.), *Slate*, and *Survival* (U.K.), and he has been featured on CNN, BBC, Al Jazeera International, National Public Radio, and Doordarshan (India).

(3) Jean-Jacques Rousseau is one of the most influential thinkers during the Enlightenment in eighteenth century Europe. In his first major philosophical work, *A Discourse on the Sciences and Arts*, (1750) Rousseau argues that the progression of the sciences and arts has caused the corruption of virtue and morality. This discourse won Rousseau fame and recognition, and it laid much of the philosophical groundwork for a second, longer work, *The Discourse on the Origin of Inequality*. One of the central claims of the work is that human

beings are basically good by nature, but corrupted by the complex historical events that resulted in (his) present day civil society.

His major work on political philosophy, *The Social Contract*, published in 1762, caused a grand brouhaha in France. Rousseau fled France and settled in Switzerland, but as might be expected, he continued to find difficulties with authorities and quarreled with friends. The end of Rousseau's life was marked in large part by his growing paranoia and his continued attempts to justify his life and his work.

(4) Professor Jared Diamond was born of Polish-Jewish heritage, to a physician father and a teacher/musician/linguist mother. He earned an AB degree from Harvard in 1958 and his Ph.D. in physiology and membrane biophysics from Cambridge University in 1961. During 1962-1966, he returned to Harvard as a Junior Fellow. He became a professor of physiology at UCLA in 1966. While in his twenties, he also developed a second, parallel, career in the ecology and evolution of New Guinea birds, and has since led numerous trips to explore New Guinea and nearby islands. In his fifties, Diamond gradually developed a third career in environmental history, becoming a professor of geography and of environmental health sciences at UCLA, his current position. Diamond speaks a dozen languages, listed in the order learned: English, Latin, French, Greek, German, Spanish, Russian, Finnish, Fore (a New Guinea language), New Melanesian, Indonesian, and Italian.

Diamond's books rely on fields as diverse as molecular biology, linguistics, physiology, and archeology, as well as knowledge about typewriter design and feudal Japan. Because of his broad expertise and the large number of articles credited to him, M. Ridley has suggested jokingly that Diamond is not a single person, but instead "is really a committee."

(5) Benny Morris "Righteous Victims: *A History of the Zionist-Arab Conflict*—1881-2001 Morris' highly acclaimed book about the Palestinian-Israeli conflict. It methodically (and in great details) walks you through the evolution of the conflict since the late 19th century, until year 2000. It should be noted that most of the early chapters of this book have been mostly based on declassified Israeli, Haganah, and Zionists documents. Author Benny Morris is a famous Israeli historian who has written several books and many articles rotating around the al-Nakba and the Palestinian refugees, especially *'The Birth of the Palestinian Refugee Problem.'* Interested in getting a detailed and relatively unbiased account of the Palestinian-Israeli conflict? then this is a must read.

(6) The conundrum of Bernie Goetz finds a partial solution in the book by Lillian Rubin. *Quiet Rage Bernie Goetz in a time of madness* New York: Farrar, Straus and Giroux. 1986.

(7) See Matt Ridley *Genome*. Ridley has a doctorate in zoology from Oxford University. A career as a journalist, has equipped him to write books about science, economics and the environment. He lives in Newcastle upon Tyne. Matt Ridley's

books have sold over half a million copies, been translated into 25 languages and been short-listed for six literary prizes.

(8) William Bratton *Turnaround: How America's Top Cop Reversed the Crime Epidemic,* New York: Random House. 1998.

(9) Research has indicated that student behavior is influenced by student dress and grooming. Consequently, student grooming is said to be the proper concern of school administrators. School staff recognizes that parents bear the primary responsibility for setting standards for their children's dress and grooming. However, because of health and safety factors, because of the influence of dress and grooming on students' attitude and behavior, and because of the need to prevent disruptive influences and preserve the academic environment of the school, student dress and grooming are proper concerns of teachers as well as administrators. DeMitchell, Todd A. Fossey, Richard and Cobb Casey. "Dress Codes in the Public Schools: Principals, Policies, and Precepts." *Journal of Law & Education* 29:1 (January 2000): 31-49. *Available from*: Journal of Law & Education, Jefferson Law Book Company, 2100 Huntingdon Ave., Baltimore, MD 21211.

(10) Judith Harris has long advocated that peers not parents are the formative factor in the ultimate outcome of a developing child. See her *The Nurture assumption.* New York. Free Press. 1998. (A very readable and newly reprinted book.)

CHAPTER 10

EDUCATION

(1) Article 2 of the first Protocol to the European Convention on Human Rights obliges all signatory parties to guarantee the right to education. At the world level, the International Covenant on Economic, Social and Cultural Rights of 1966 guarantees this right under its Article 13.

(2) Robert Frost quoted in the United Press International . . . reprinted in The Week, Feb. 6, 2009. Frost, considered one of America's premier poets, had a difficult life to say the least. Those who only know of Frost through his poetry should read through some of his biographical information. I believe it will expand understanding of his poetry. For example, he was born 1874 in San Francisco, and named after Confederate General Robert E. Lee. In 1883—Frost allegedly hears voices when left alone and is told by mother that he shares her gift for "second hearing" and "second sight." His father, a newspaper editor drinks himself to death. (Died in 1885) In 1892, Frost becomes engaged to Elinor White (a classmate)

but he is dependent upon grandparents for financial support, enters Dartmouth College instead of Harvard because it is cheaper, and because grandparents blamed Harvard for his father's bad habits. Bored, he drops out of Dartmouth at the end of December. In 1894 He returns to teaching grades one through six in Salem. In March, he learns <u>The Independent</u> will publish his poem "My Butterfly: An Elegy" and will pay him $15. He tries to convince Elinor to marry him by getting a printer to make him two copies of a collection of his poems, called Twilight. Goes to visit Elinor to present her with a copy, but is thrown into despair by her cool reception; destroys his own copy and returns home—depressed. But in 1895 Frost starts teaching at Salem district school and Elinor White finally agrees to marry him. On December 19, 1896 a *Swedenborgian* minister marries them and nine months later—Son Elliott is born on September 25.1897—he then Passes Harvard College entrance examinations, borrows money from grandfather and enters Harvard as a freshman. 1899—and then again becomes bored/and/or restless and drops out of Harvard on March 31. Daughter Lesley is born on April 28. Mother has advanced cancer and dies the following year 1900). His life goes on with tragedy after tragedy but he lived to almost ninety (died 1963) after winning four Pulitzer Prizes and seventeen honorary degrees.

(3) Ponzi schemes are a type of illegal pyramid scheme named for Charles Ponzi, who duped thousands of New England residents into investing in a postage stamp speculation scheme back in the 1920s. Ponzi thought he could take advantage of

the differences between U.S. and foreign currencies used to buy and sell international mail coupons. Ponzi told investors that he could provide a 40% return in just 90 days compared with 5% for bank savings accounts. Ponzi was deluged with funds from investors, taking in $1 million during one three-hour period—and this was 1921! Though a few early investors were paid off to make the scheme look legitimate, an investigation found that Ponzi had only purchased about $30 worth of the international mail coupons. Decades later, the Ponzi scheme continues to work on the "rob-Peter-to-pay-Paul" principle, as money from new investors is used to pay off earlier investors until the whole scheme collapses.

(4) In the 1960s, William F Buckley, Jr., wrote: "The Beatles are not merely awful, I would consider it sacrilegious to say anything less than that they are god-awful . . . They are so unbelievably horrible, so appallingly unmusical, so dogmatically insensitive to the magic of the art, that they qualify as crowned heads of anti-music."

However, the early 60s, for some, were a thrilling innovative time. It was, for example, exciting to see all the new stars on American Bandstand! One after another . . . all the biggies of rock and roll came on the show! Little Richard; Jerry Lee Lewis; Buddy Holly; Connie Francis; the Platters; the Drifters; Chuck Berry; the Coasters; Fats Domino; somebody named Elvis-somebody-or-another . . . the list was endless! There were, of course, no oldies at that time . . . as these singers, songs, songwriters and performers were the main ingredients of the beginning of the rock and roll era! But, everyone did not like rock and roll at the time! As a matter of fact, it was downright hated! Parents; school officials; the government; churches; and the "establishment," frowned on rock and roll right from the beginning! "Rock and roll was corrupting our kids!" Was the cry! Certain songs were banned from church, and the "establishment frowned on rock and roll. It was later in the mid-1960's, when the drug culture came onto the scene, and it was then . . . rock and roll was more accepted by the establishment!" After all, parents thought it was best for their kids to be having fun with music . . . than to get into the hazards of drugs!

If you were over 30 years old at that time, you were OLD . . . not trusted . . . and you were part of the "establishment!" You were, what Bill Cosby calls, "OLD PEOPLE!"

(5) Thomas Barnett's book, *Great Powers, America and the World after Bush* revealed him to be one of the most important strategists of modern times. In his last book *The Pentagon's New Map: War and Peace in the Twenty-First Century,* Barnett drew on a fascinating combination of economic, political and cultural factors to predict and explain the nature of modern warfare. He presented concrete, world-changing strategies for transforming the US military into a two-tiered power capable not only of winning battles, but of promoting and preserving international peace. You can see him lecture on the web at *www.TED.com*. Dr. Barnett has been a senior adviser to military and civilian leaders in a range of offices, including the

Office of the Secretary of Defense, the Joint Staff, Central Command and Special Operations Command. During the period from November 2001 to June 2003, he advised the Pentagon on transforming military capabilities to meet future threats. He led the five-year New-Rule-Set-Project, which studied how globalization is transforming warfare. The study found, among other things, that when a country's per-capita income rises above ~$3,000—war becomes much less likely.

(6) Education is today largely paid for and almost entirely administered by governmental bodies or non-profit institutions. This situation has developed gradually and is now taken so much for granted that little explicit attention is any longer directed to the reasons for the special treatment of education even in countries that are predominantly free enterprise in organization and philosophy. The result has been an indiscriminate extension of governmental (bureaucratic) responsibility. A stable and democratic society is impossible without widespread acceptance of some common set of values and without a minimum degree of literacy and knowledge on the part of most citizens. Education contributes to both. In consequence, the gain from the education of a child accrues not only to the child or to his parents but also to other members of the society; so that requiring that each child receive a minimum amount of education of a specific (*nodulized*) kind is required. Such a requirement could be imposed upon the parents without further government action, just as owners of buildings and frequently of automobiles, are required to adhere to specific standards to protect the safety

of others. One argument for *nationalizing education* is that it might otherwise be impossible to provide the common core of values deemed requisite for *(brainwashed) social* stability. The imposition of minimum standards on privately conducted schools might not be enough to achieve this result. The issue can be illustrated concretely in terms of schools run by religious groups. Schools run by different religious groups will, it can be argued, instill sets of values that are inconsistent with one another and with those instilled in other schools; in this way they convert education into a divisive rather than a unifying force.

(7) Thomas Jefferson wrote a great deal about education. The following quotes, more or less, sum up his general attitude about educating the masses in "new America".

"At every of these schools shall be taught reading, writing and common arithmetick, and the books which shall be used therein for instructing the children to read shall be such as will at the same time make them acquainted with Graecian, Roman, English and American history. At these schools all the free children, male and female, resident within the respective hundred, shall be entitled to receive tuition gratis, for the term of three years, and as much longer, at their private expense, as their parents, guardians or friends, shall think proper."

"The value of science to a republican people, the security it gives to liberty by enlightening the minds of its citizens, the protection it affords . . . in short, its identification with power, morals, order and happiness (which merits to it premiums of encouragement rather than [repression])"

Thomas Jefferson: On the Book Duty, 1821.

(8) "Students in Singapore and several other Asian countries significantly outperform American students, even those in high-achieving states like Massachusetts, the study found." Asian economic competitors are winning the race to prepare students in math and science," said the study's author, Gary W. Phillips, chief scientist at the American Institutes of Research, a nonprofit independent scientific research firm.

(9) T. L. Friedman's provocative books and journalistic contributions have probably done more than any other single thing to change snobby America's attitude toward the conservation (green) movement in America at large. He is certainly one of the most thoughtful and original thinkers of our times. Born in Minneapolis on July 20, 1953, Mr. Friedman received a B.A. degree in Mediterranean studies from Brandeis University in 1975. In 1978 he received a Master of Philosophy degree in Modern Middle East studies from Oxford. Mr. Friedman is married and has two daughters. He won the 2002 Pulitzer Prize for his commentary, and his third Pulitzer for The New York Times. He became the paper's foreign-affairs columnist in 1995. *"The World is Flat: A Brief History of the 21st Century,"* was released in April 2005 and won the inaugural Goldman Sachs/Financial Times Business Book of the Year award. His book, *"From Beirut to Jerusalem"*

(1989), won the National Book Award for non-fiction in 1989 and "The Lexus and the Olive Tree" (2000) won the 2000 Overseas Press Club award for best nonfiction book on foreign policy and has been published in 27 languages. Mr. Friedman also wrote *"Longitudes and Attitudes: The World in the Age of Terrorism*" (2002) and the text accompanying Micha Bar-Am's. "Israel: A Photo-biography". His most recent book, *Hot, Flat and Crowded,* concerns the HOW to provoke a major "green" revolution in America and HOW this alone can provide economic stability.

(10) The term "conservation ethic" is already in play within the younger generation. Like electronic competency, *conservation* is part and parcel of "youth consciousness." Just as children "shamed" their parents into quitting smoking, they are now energetically pursuing similar tactics to bring about universal conservation and re-cycling.

(11) The *Greatest Generation* is a term coined by journalist Tom Brokaw to describe the generation of Americans who grew up during the Great Depression and then fought to victory in World War II, as well as those who supported WWII on the home front by making decisive material and motivational sacrifices—all of which contributed to the war effort. After surviving the war, they then went on to build and rebuild United States industries so that America became the superpower of the era. The *Greatest Generation* follows the *Lost Generation* of the 1920s and precedes the *Silent Generation* of the 1950s. Of course, it seems likely that some social/historian

from the *Boomer Generation* will one day write the definitive argument against the nomenclature (greatest) being applied to anything.

(12) Leonard Cohen. <u>*Anthem.*</u> The line "a crack in everything" may come from a book by Jack Kornfield on Buddhism. The story is that a young man who had lost his leg came to a Buddhist monastery and was extremely angry at life. He constantly draws pictures of cracked vases, because he felt damaged. Over time, he found inner peace, and changed, but still drew broken vases. His master asked him one day: "Why do you still draw a crack in the vases you draw, are you not whole?" And he replied "yes, and so are the vases. The crack is how the light gets in"

CHAPTER 12

ADDICTION

(1.) It is, however, asserted that polygamy and the seclusion of women, as enjoined in the Holy Qur'an, have done more harm to a woman than the benefit conferred on her by bestowal of property rights. The fact is that a great misunderstanding exists on these two points. <u>Monogamy is the rule in Islam</u> and polygamy only an exception allowed subject to certain conditions. The following two verses are the only authority for the sanction of polygamy, and let us see how far they carry us:

"And if you fear that you cannot act equitably towards orphans, marry such women as seem good to you, two and three and four; but if you fear that you will not do justice between them, then marry only one or what your right hands possess: this is more proper that you may not deviate from the right course" *(4:3).*

"And they ask thee a decision about women. Say, Allah makes known to you His decision concerning them, and that which is

recited to you in the Book concerning orphans of the women to whom you do not give what is appointed for them while you are disinclined to marry them" (4:127).

Now the first of these verses allows polygamy on the express condition that "you cannot act equitably (financially) towards orphans", and what is meant is made clear by the second verse, which contains a clear reference to the first verse in the words, "that which is recited to you in the Book concerning orphans of women". Some Muslims were guilty of a double injustice to widows: they did not give them and their children a share in the inheritance of their husbands, nor were they inclined to marry widows who had children, because the responsibility for the maintenance of the children would in that case devolve upon them. The Qur'an remedied both these evils; it gave a share of inheritance to the widow with a share also for the orphans, and it commended the taking of such widows in marriage, and allowed polygamy expressly for this purpose. It should, therefore, be clearly understood that monogamy is the rule in Islam and polygamy is allowed only as a widow and her children. This permission was given at a time when the wars, which were forced on the Muslims, had decimated the men, so that many widows and orphans were left for whom it was necessary to provide. A provision was made in the form of polygamy so that the widow should find a home and protector and the orphans should have paternal care and affection. It appears that Europe today has an excess of women, and let it consider if it can solve that problem otherwise than by sanctioning a limited polygamy. Perhaps the only other way is prostitution, which prevails widely in all European countries yet where the law of the country does not recognize it.

(2.) Renouncing Islam can carry the death penalty in a number of countries including Iran, Saudi-Arabia, Afghanistan, Pakistan, Sudan and Mauritania. In other countries courts don't punish people who turn their backs on their faith, but family and friends often ostracize them. It's a difficult subject among Muslim communities in Europe too. See www.Islam-Watch.org

(3) According to cognitive dissonance theory, there is a tendency for individuals to seek consistency among their cognitions (i.e., beliefs, opinions). When there is an inconsistency between attitudes or behaviors (dissonance), something must change to eliminate the dissonance. In the case of a discrepancy between attitudes and behavior, it is most likely that the attitude will change to accommodate the behavior.

Two factors affect the strength of the dissonance: the number of dissonant beliefs, and the importance attached to each belief. There are three ways to eliminate dissonance: (1) reduce the importance of the dissonant beliefs, (2) add more consonant beliefs that outweigh the dissonant beliefs, or (3) change the dissonant beliefs so that they are no longer inconsistent.

Dissonance occurs most often in situations where an individual must choose between two incompatible beliefs or actions. The greatest dissonance is created when the two alternatives are equally attractive. Furthermore, attitude change is more likely in the direction of less incentive since this results in lower dissonance. In this respect, dissonance theory is contradictory to most behavioral theories, which would predict greater attitude change with increased incentive (i.e., reinforcement).

RESOURCES & RELEVANT READINGS

Adams, Douglas. 1979. The *Hitchhiker's Guide to the Galaxy.* New York: Pocket Books.

Adams, Scott. 2001. *God's Debris: A Thought Experiment.* Kansas City: Andrews McMeel—Publishing.

Alexander, Karen. 2011. *Twelve Steps to The Compassionate Life.* New York: Alfred Knopf

Alper, Matthew. 1996. *The "God" Part of the Brain.* New York: Rogue Press.

Asimov, Isaac. 1972. *The Gods Themselves.* New York: Bantam Books.

Baker, Nicholson. 2006. *Human Smoke, The Beginnings of World War II, the End of Civilization.*

New York: Simon & Schuster.

Barrow, John D. 2000. *The Universe That Discovered Itself.* New York: Oxford Univ. Press.

Brand, C. 1981. *Personality and Political Attitudes.* In Dimensions of Personality; Papers in Honour of H. Eysenck, ed. R. Lynn. Oxford: Pergamum Press. 7-38, 28.

Bryson, Bill. 1990. *Mother Tongue:* New York: Perennial Press.

Chiu, Lisa, Seachrist. 2005. *When a Gene Makes You Smell Like a Fish.* New York: Oxford University Press.

Chown, Marcus. 2002. *The Universe Next Door: The Making of Tomorrow's Science.* New—York: Oxford University Press.

Costa, Rebecca D. 2010. *The Watchman's Rattle. Thinking Our Way Out of Extinction.* New York: Vanguard Press. Perseus Book Group.

Croswell, Ken. 2001. The *Universe of Midnight: Observations Illuminating the Cosmos.*—New York: The Free Press.

d'Aquili, Eugene and Andrew B. Newberg, 1999. *The Mystical Mind: Probing the Biology of—Religious Experience (Theology and the Sciences).* Minneapolis: Fortress Press.

Davies, Kevin. 2001. *Cracking the Genome: Inside the Race to Unlock Human DNA.* New—York: The Free Press.

Dawkins, Richard.1998. *Unweaving the Rainbow: Science, Delusion and the Appetite for—Wonder.* Boston: Houghton Mifflin Company.

Dawkins, Richard. 1999. *The Extended Phenotype: The Long Reach of the Gene.* Oxford:—Oxford University Press.

Dennett, Daniel C. 1995. *Darwin's Dangerous Idea: Evolution and the Meaning of Life.*—New York: Touchstone.

Diamond, Jared. 2004. *Collapse: How societies choose to fail or succeed.* New York: Viking Press.

Dugatkin, Lee.1999. *Cheating Monkeys and Citizen Bees: The Nature of Cooperation in—Animals and Humans.* New York: The Free Press.

Edelen, William. 1998. *Spirit Dance Essays.* Boise: Joslyn & Morris Printing.

Edelen, William. 2000. *Earthrise Essays.* Sunday Symposium Palm Springs.

Edelman, Gerald. 1992. *Bright Air, Brilliant Fire.* New York: Basic Books.

Ehrsson, Hennk. 2004. *Facilitated Belief.* Science on Line. July 2004.

Einstein, Albert. 1954. *Ideas and Opinions*. New York: Bonanza Books.

Erikson, Erik. 1950. *Childhood and Society*. New York: W.W. Norton & Co.

Evatt, Cris. 2010. *The Myth of Free Will*. Sausalito. California. Café Essays.

Friedman, Thomas. 2005. *The World is Flat*. New York: Farrar, Straus and Giroux.

Gardner, Martin. 1957. *Fads & Fallacies In the Name of Science*. New York: Dover—Publications.

Giovannoli, Joseph. 1999. *The Biology of Belief: How Our Biology Biases Our Beliefs and—Perceptions*. New York: Rosetta Press.

Gladwell, Malcolm. 2005. *Blink. The Power of Thinking Without Thinking*. New York: Little Brown Publishers.

Gladwell, Malcolm. 2008. *Outliers. The Story of Success*. New York: Little Brown Publishers

Goldsmith, Donald. 2000. *The Runaway Universe*. Cambridge, Mass.: Perseus Books.

Gould, Stephen Jay. 1996. *Full House: The Spread of Excellence from Plato to Darwin*.—New York: Harmony Books.

Green, Brian. 1999. *The Elegant Universe: Superstrings, Hidden Dimensions, and the Quest—for the Ultimate Theory*. New York: W.W. Norton.

Greene, Brian. 2004. *The Fabric of the Cosmos: Space, Time, and the Texture of Reality*.—New York: Alfred A. Knoff.

Greenwood, Michael and Peter Nunn. 1994. *Paradox & Healing: Medicine, Mythology &—Transformation*. Victoria, B.C.: Paradox Publishers.

Hamer, Dean. 2004. *The God Gene: How Faith is Hardwired into Our Genes*. New York:—Doubleday.

Hampden-Turner, Charles. 1981. *Maps of the Mind*. New York: Macmillan Publishing Co.,—Inc.

Harris, Sam. 2005. *The End of Faith*. New York: W.W. Norton and company.

Harris, Sam. 2010. *The Moral Landscape*. New York: Free Press.

Harris, Judith. 1998. *The Nurture Assumption: Why Children Turn Out the Way They Do.*—New York: Simon and Schuster. Touchstone.

Hawking, Stephen W., Kip S. Thorne, Igor Novikov, Timothy Ferris, and Alan Lightman. 2002. *The Future of Space-time*. New York: W.W. Norton.

Hawking, Stephen. 1996. *A Brief History of Time*: The Updated and Expanded Tenth—Anniversary Edition. New York: Bantam Books.

Hofstadter, Douglas. 2007. *I Am a Strange Loop:* New York: Basic Books.

Holmes, Ernest. 1998. *The Science of Mind: A Philosophy, A Faith, A way of Life:* New York: Penguin Putman Inc.

Iles, Greg. 2003. *The Footprints of God*. New York: Pocket Star Books.

Jaynes, Julian. 1976. *The Origin of Consciousness in the Breakdown of the Bicameral—Mind*. Boston: Houghton Mifflin Company.

Jensen, Derrick. 2006. *Endgame. The problem of civilization*. New York: Seven Stories Press.

Kaku, Michio. 2005. *Parallel Worlds: A Journey Through Creation, Higher Dimensions, and—the Future of the Cosmos*. New York: Doubleday.

Lamott, Anne. 2005. *Plan B Further Thoughts on Faith*. New York: Riverhead Books.

Milton, Robert. 2004. *The Unspoken*. Bloomington, Indiana: Author House.

Milton, Robert. 2005. *The True Believers: The Golden age of Terrorism*. Bloomington Indiana: Author House.

Muller, Richard. A. 2008. *Physics for Future Presidents.* New York: W.W. Norton.

Moyers, Bill. 1995. *Healing and the Mind.* New York: Broadway Books

Newberg, Andrew, Eugene D'Aquili and Vince Rause. 2001. *Why God Won't Go Away.* New York: Ballantine Books.

Nicolelis, Miguel. 2005. The Cortical Ensemble Adaptation. *J. of Neuroscience.* 25:4681-4693.

Park, Robert. 2000. *Voodoo Science: The Road from Foolishness to Fraud.* New York:—Oxford University Press.

Pinker, Steven. 1997. *How the Mind Works.* New York: W.W. Norton & Company.

Pinker, Steven. 2002. *The Blank Slate: The Modern Denial of Human Nature.* New York: Penguin Books.

Quinn, Daniel. 1992. Ishmael: *An Adventure of the Mind and Spirit.* New York: Bantam/Turner Books.

Ridley, Matt. 2000. *Genome.* New York: Harper Collins Publishers.

Ridley, Matt. 1993. *The Red Queen: Sex and the Evolution of Human Nature.* London: Penguin Books.

Rucker, Rudy. 1984. *The Fourth Dimension: A Guided Tour of the Higher Universe.* Boston: Houghton Mifflin Company.

Rymer, Russ. 1993. *Genie: A Scientific Tragedy.* New York: Harper Perennial.

Sagan, Carl and Ann Druyan. 1992. *Shadows of Forgotten Ancestors.* New York: Random House.

Sagan, Carl. 2000. *Carl Sagan's Cosmic Connection.* New York: Cambridge University Press.

Sheldrake, Rupert. 1988. *The Presence of the Past: Morphic Resonance & the Habits of Nature.* Rochester, Vermont: Park Street Press.

Shermer, Michael. 2002. *Why People Believe Weird Things*. New York: Holt and Company.

Sullivan, Harry Stack. 1953. *The Interpersonal Theory of Psychiatry*. New York: W.W.—Norton.

Weil, Andrew and Winifred Rosen. 2004. *From Chocolate to Morphine: Everything You—Need to Know about Mind-Altering Drugs*. Boston: Houghton Mifflin Company.

Weil, Andrew. 1998. *The Natural Mind*. Boston: Houghton Mifflin Company.

Weinberg, Steve. 2001. *Facing Up: Science and Its Cultural Adversaries*. Cambridge, Mass.: Harvard University Press.

INDEX

4:20 115-16
9/11 70, 74, 150, 159, 206, 268, 330
9/11 bombers 159

A

absurdities 10, 74, 80, 83, 87, 151, 160, 162, 296, 298
absurdity 61, 64-5, 70-1, 74, 78, 83, 86, 162, 182-3
acupuncture 138, 141
adaptive brain-systems 77
addiction to future abstractions 255
addictive state 249
Affective Communications Test 123
Afghanistan 152, 154, 237, 241, 270, 356
African mistress 207
agape 163
Age of the Renaissance 110
aggression 62
AIDS 58
Air Force One 23
Alexander Graham Bell 117, 319
Allah 31, 64, 84, 90, 155, 178, 337, 354
Allah in action 155
Allah's will 64

alternate realities 48
alternate reality 22
alternative for childcare 220
American dream 185, 187, 192, 229
American moderates 149
Americanized brain nodules 150
America's ruin 202
America's success 72, 191, 202
Amygdala 62
ancient intelligence 69-71
Ancient Rome 243
anti-Catholic 181
anti-war identities 239
Arabian poetic literature 17
Aristotle 55-7, 90, 262
arm-twisting 36
Armageddon 155, 166
Arms control 243
Armstrong 53, 169, 175, 263-4, 337-8
Arnold Schwarzenegger 32-3
atheists 58
atomic weapons 197
Audacity of Hope 225
Australian Outback 69
avenging angel of NY 198

B

baby-sitting facility 220
Bach 39-40

bailout 213, 224
bankruptcy 46, 232, 234, 324
baptism 181
bear 64, 140, 232, 292, 344
Beelzebub 84
Beijing 73
belief addictions 127
believe uncompromisingly 113
benign representatives of Islam 150
Benny Morris 197, 343
Bernie Goetz 198, 343
Bertrand Russell 215, 267
bi-lingual populations 214
Big Bang 131, 266, 326
big education 210
big government 210
birth of a new century 73
black holes 57, 106
blasphemy 180, 338-9
Bloom, and Lazerson 27
Bonobos 95, 104
Boogie Woogie 217
Boomers 214
Boston Globe 151
Brahma 84
brain cloud 159
brain development 87, 259
brain plasticity x-xi, 8, 15, 22, 283-4, 296
brain scans 11, 24, 83, 131, 252, 269, 281
Brain Spas 8
brains physically change 25
brainwashing 163
British Broadcasting Company 180
Broken Window Syndrome 199
Bruno 34, 97, 269
BS (bad science) detector 67
bubble of laughter 37
Buddha 13, 16, 175, 216, 265
Buddhist 28, 31, 328, 353
Buddhist Priests 28
Business as usual 226

C

Cambodia 55
Candice Pert 94, 310
Cantata 39
Captain Cook 75
Caribou 81
Carl Sagan 67, 363
Catholic Nuns 28
Celestine Prophecy 126
cell assembly 26
chakra 17
Champagne 65
characteristic 3, 76, 295
Charlie Rose 158
charter-school 215
chauvinistic pride 112
chi 17
Chicago Tribune 151
China 73-4, 111-12, 141, 188-90, 194, 240
China's citizens 194
Chris Brand 49
Christ 13, 309
Christian Bible 100, 180, 254, 337
Christian world 9, 157
Christmas 50, 137
Christopher Columbus 165
civilization's end 161
clairvoyance 114
Clever Hans 124
clinical depression 248
cocaine addiction 251-2
coincidence upon coincidence 121
coitus 253
Communism 74
complex form of thinking 219
compromise can mitigate 258
compulsive 250
computer/electronic skills 217
computer literate generation 169
conceit 154
condominium associations 79
Conflict and cooperation 78
Confucius 176

contempt for legal precedent 233
Copernicus 95, 106
correlation does not equal causation 65
cosmic joke 11, 83, 87
courteous concessions 162, 182
crime in NY 199
crippling cynicism 109
Cris, Robin and Whoopi 70
Crusades 72
cry-wolf memorandum 212
cults of positivity 45
cultural micro-rhythms 123, 322
cum hoc ergo propter hoc 65
current cultural contour 218
cynical 109

D

Dale Carnegie 122
Damasio 129, 325
dangerous science 108
Daniel Pearl 151
Darwin 93-5, 104, 106, 289, 307-8, 312, 360-1
Dave Crisp 55
Dawkins 40-1, 53, 289, 360
DC economic professionals 215
death 21-2, 47-8, 81, 84, 101, 113, 121, 176, 180, 182, 229, 257, 264, 270, 339
death by devil 182
Declaration of Independence 207
delusion 75, 244, 289, 323, 360
demise anxiety 22
Diamond 80-1, 195-8, 303, 342-3, 360
different color face 62
dig ditches 205
dog Dugan 55, 159, 287
Don Jordon 277
Dopamine 248-9
double-checks of logic 64
Dr. Frankenstein 100, 102

dualism 21-2, 44, 120, 129, 133-4, 137, 139, 263-4
dualistic believer's conclusion 133
dualistic discord 134
Duke University 140
Dupe Gene 5
dupe gene complex 42, 47
dupe nodules 49
duped 5, 12, 46-7, 49, 51-2, 62-3, 68, 89, 295, 346
Dwight D. Eisenhower 231
dysfunctional neighborhood 202
dysfunctional relatives 201

E

Easter bunny 50, 107
Easter Island 80, 82
eclipse 93-4
ecumenical movement 181
education 25, 27, 49-50, 62, 154, 168, 205-6, 208-13, 221, 223-4, 236, 307-8, 331-2, 344-5, 349-50
education available to every citizen 208
Education Week 236
educational fantasy 208
educational problem 214
Egypt 149, 178, 329
Eleanor Porter xi, 42
electronic comfort 216
empathetic 143, 183
empathetic 'therapists'. 183
ends justify the means 234
epicenter of rational thinking 87
epinephrine 247-8
erased the graffiti 201
Eric Lander 104, 314
Eric Nestler 252
Ernest Holmes 13, 15, 27, 45, 130, 302
eros 163
eternal hell 69
Ethanol 75-6

ethical effort 193
Europe 111, 179, 238-9, 242, 309, 316, 329, 341, 355-6
European Nations 188
European thought 110
evolution 14, 23, 66, 69-71, 74, 77, 89-92, 94, 97, 104, 106, 177, 225, 307-8, 342-3
Evolutionary biologists 75
evolved civility 196
experimental bent 6
explanation for human absurdity 83
exploitation 80, 152, 213, 330, 334

F

fair and balanced 79, 246
faith 17, 64, 76, 81, 86, 94, 127, 138, 158-62, 169-70, 182, 257, 267-8, 276, 361-2
false-positives 141
flight and fright 258
floundering syndrome 235
fMRI 6, 11, 22, 129-30, 169, 252, 285
food bingeing 24
foot washing 181
Ford 224
Forer Effect 4
Founding Fathers 91, 167, 207, 262, 336
France 23, 178, 240-1, 283, 342
free education 209, 212
free educational system 224
Free speech 155
free will 21, 25, 31-2, 107, 143, 199, 268-9, 277, 361
Frenchmen 21, 23, 26, 64
Friction 163
frontal lobe damage 252
frontal lobes 28, 85, 249-51
full moon 63

G

galaxies 96, 145
Galileo 34, 93-5, 97-9, 106, 131, 145, 266, 269, 311
Gay gene 42, 293
genes 3, 42, 48, 51, 101, 103, 142, 199, 290-4, 314-15, 361
genius 39-40, 116-18, 206
Genome 101, 290, 296, 300, 315, 343, 360, 363
genome-based revolution 7
genomic sequencing 105
genotype 22, 47, 292
George Seurat 117
giant quantum field 115, 317
giant redwoods 115
giggly optimism 109
GJ Wang 251
God gene 42, 293, 361
God's will 171, 254
Goldilocks zone 95
Goldwater 177
Google x, 134, 168, 180, 267, 294, 296, 331, 333, 336
GPS 56-7
Great Depression 210, 352
Greenland 81
Greenland Viking 81
gridlock 61, 70, 304
ground zero 150, 330
group drumming 84
Gulf States 190
Guru 18, 63, 78, 124-5, 164
Gustave-Roussy Institute 102, 313
Gutenberg 110, 218
GW Bush 243
Gypsy moth 246

H

habituated brain nodules 230
Hans 66, 124
Hans the horse 66
harbor of consciousness 14, 22, 31, 54, 145, 267, 278

hardwired to believe 183
harem 253
Harris 53, 259, 327, 344, 362
Hawaii 75, 96, 189
Hebb 15, 25-6, 259, 282-3
Hebrew Bible 254
herd animals 34
here and now 12, 128, 246, 256
heroic dreams 235
high tech intelligence 69
hippocampus 24, 281-2, 292
history of Europe 238
Hitchens 53
Holland 178, 241
Hollywood films 191
Hollywood movies 187
Holmes 13, 15, 25, 27, 45, 130, 259, 262, 288, 302, 362
Holy Book literalism 179
Homo sapiens 48, 58, 72, 90, 93, 101, 104, 131, 137, 262, 321
hope 46, 51-3, 77, 93-4, 178, 213, 216, 224-7, 242, 266, 276, 295-6, 301, 329, 337
Horatio 64, 313
human sexuality 247-8
humanists 142
hundredth-monkey phenomenon 126
hybrids 224

I

identity crisis 219
ignorance is bliss 31
ignorant nation 225
illegal occupants 185
impaired judgment 250
impulsive 250
India 18, 75, 112, 188-90, 332, 341
Indian entrepreneurial triumph 191
Indrek Wichman 177
infallible utterances 168
inspired writings 83, 168, 178, 255

Intelligent adaptability 68
Intelligent Design 90, 93
Internet iv, x, 112, 218, 247, 267
intractable silliness 70
intuition 67, 125, 136
intuitively gifted student 141
inventiveness 116
Iran 158, 356
Iraq 152, 154, 186, 237, 270
Islam Imams 156, 332
Islamophobic 241

J

James Madison 223, 309
James Randi 125
Japan 141, 189, 343
JD Unwin 81
Jeff Schwartz 35
Jesus 103, 158, 161, 175, 338
Jewish nanny 126
Jewish religion 170
J.F. Kennedy 232
Jihad 160
Jostein Gaarder 114
Judaism 171, 338
Junk 103
Jurassic park 103
justifiable war 240

K

Kabala 17
Karen Armstrong 53, 175, 263, 337
Keats 41, 131
killing products 232
kingdom of God is within 57
the knowledge 24
Kodak Center 63
Kool-Aid path 85
Korea 152, 312
Korean conflict 163
kosher concepts 170
Kundera 121, 321

L

La Brea tar pits 101
lame duck 221
large rewards 114
Las Vegas 40, 71, 306, 320, 332
Law of Parsimony 80, 84, 130
Lehrer 141
Leonard Cohen 10, 230, 353
lesser-qualified men 253
liberal/intellectualism 179
libido 249
limbic system 62
little bit pregnant 165, 184
live well—love much—laugh often 128
lobbying 210, 232-3
London cab drivers 24
love of magic 50
Love of self 163
love that soldiers have for each other 236
LSD 30
luck 53-6, 67, 192
lunacy 155-6, 160

M

M. Stryker 28
MAC computer 220
Madison Avenue 185, 191
Madrasah education 154
Magic Flute 117
magic kingdom 51
man-made chromosome 99
Manifest Destiny 186
Mary Baker Eddy 45
Mayans 81
meanness and war 79
media 70, 73, 78, 150-1, 161-2, 182, 185, 228, 237, 289, 319
Media facilitated the greatest generation 228
mediocrity 82, 227, 304
melting pot 152
memory 8, 15, 89, 122, 256, 283, 292, 300-1, 320
ménage a trios 245
Mencken 57, 288, 297
mental cuisine 23
mental working-out 27
Merton 116, 118, 318-19
messianic demagogue 239, 242
methamphetamine 251-2
micro-movements 123
microscope 71
Microsoft 220
Mid-East Theocracies 36
Mike Kinsley 186
military industrial complex 231-2
missionaries 110-13, 181
Mistakes 193
MIT graduate 85
Mo'ai rock statues 80
moderate secularists 242
moderation 149-50, 152-3, 156, 161, 163, 181, 184, 329
Modern urban legends 189
Modernists 9
modifiable brainpower 72
Mohamed 175
mom training 222
Mongolia 74
Mongolian plains 69
mood swings 250
moral calculus 240
Moses 175-6
Mother Nature 72
motor mimicry 122
Mount Cedar Sinai in New York 252
Mozart 117
Mr. Dugan 55, 159
multi-ethnic societies 61
multi-tasking mothers 50
Multiple orgasms 253
multiple spouses 253
murder 156, 178, 271, 300-1
music 39-40, 117, 216-17, 288, 347-8
Musicology 39

Musicophilia 85, 306
Muslim 9, 17, 25, 31, 154-8, 160-1, 165, 177-80, 184, 240-1, 253, 255, 270, 277, 331-2
Muslim world 9, 157, 253, 331
mystical 4, 11, 13, 29, 39-40, 83-5, 120, 130, 319, 360
Mystical understanding 4
myth of OZ 190
mythical ideology 73

N

narcotic addiction 249
NASA 95-6
National Academy of Sciences 251, 282, 302
nationalistic chaos 111
Nations of the Continent 239
Native Americans 186, 236
natural addiction 251
naturalist 125, 277
Naturalists 59
Nature 44, 57, 59, 72, 252, 282, 301, 360, 363
Nature Neuroscience 252
nature vs. nurture 41
Nazis 230
near death 84
Neuro-scanning research 137
New Age' spiritual systems 113
new CD 129
New Deal 210
New Guinea Highlands 195, 197
new kind of invasion 242
New Scientist 93, 309
New Thought 45-6, 129-30, 132
New Yorker 138, 141
Newberg 28, 31, 84, 129, 305, 315, 325, 335, 360, 363
Newton 41, 93, 131, 306
next big thing 127, 227
Nicholson Baker 234
Nirvana 30
Nobel Peace Prize 149

nodule xi, 11, 14, 16, 26-7, 43-4, 47-8, 58, 76-7, 82, 111, 153, 211-12, 283, 294-5
nodules 10-11, 15-16, 19, 26-8, 30-3, 35-6, 77-8, 126-8, 149-50, 154-5, 158-60, 201-3, 258-60, 265, 267-9
nodules triumphed 159
nodulized biases 76
Noetic science 130
nuclear fission 161
nucleus accumbens 249
numerology 144, 182

O

Obama 222, 295, 329-30
Occam's razor 84, 109, 126, 306
old guard 216
old-wives-tales 165
Oliver Sacks 85, 306
Olympian gods 49
Olympic champions 222
One x, 7, 17, 23, 29-30, 64, 85, 94, 98, 123, 175, 181, 198, 200, 237
ONE substantial whole 139
orgy 245, 253
orientation/ association area 28
origami feline 153
Orthos (straight ones) 171
Oskar Pfunger 124
otherworldly 121

P

Pacific islanders 69
Pakistan 180, 339, 356
pan-theism 84
paper tiger 149, 153
parable for us today 82
pathologic gambling 252
patriotism 111-12, 167, 222, 243
Paul Butler 95
peaceful Muslims 157
peer pressure 171

Penn & Teller 86
perfect storm 161
Peyote 30
phase sequence 26
philosophical cant 6
philosophical mentors 5
Phineas Quimby 45
Plan A 212
plastic human brain 111
playground bully 165
pleasure pathways 249
pleasure thermostat 249
Pollyanna xi-xii, 4-5, 12, 19, 22, 35, 40, 42-4, 47-52, 67, 69, 122, 187, 294
Pollyanna believer 52
Pollyanna Brain 19
Pollyanna brain in action 122
Pollyanna complex 22, 35, 43
Pollyanna gene 42, 47, 49
Pollyanna gene complex 42, 47, 49
Pollyanna nodule 44
Pollyanna temperament 49
Pollyanna thinking 51
Pollyannaish ideas 63
Ponzi Scheme 213
Pope's Astronomer 97
pornography 114, 246-7, 250
positive thinking vs. negative thinking 45
Power of Positive Thinking 47
pre-frontal area 29
prefrontal anticipatory faculties 70
prefrontal lobes 47, 248
primitive crime 240
prognosticators 45, 109, 121
Prophets' benevolent Quran message 255
psychedelic experience 30
psychological research 5

Q

quack science 67
quantum jargon 116
Quantum physics 57, 113, 115, 119, 318
Quran 156, 179, 254-5, 331, 337

R

Rabbis 170, 337
rainbow 40-1, 131, 360
Ralph W. Emerson 44
rat moms 221
rat pups 221
rational focus 162, 183
reality check 233
Rebecca Costa 81, 304
reductionism 134, 144
regression to the mean 67
REHAB xi, 14, 24, 27, 34, 36, 58, 78, 127-8, 183, 203, 243, 252, 257-8
Religious Science 13, 132, 302
revolutionary information 8
Richard Dawkins 40, 53, 289
Richard Ebright 102
Richter 51-2, 296
Rick Warren 158
Rick Weiss 105, 312
Robert Merton 116, 318
Robert Sapolsky 62
Romeo and Juliet 117
Rousseau 59, 194, 341-2

S

sacrilegious 16, 29, 347
Salieri 117
Sanskrit 176
Santa Claus 48, 50, 265, 277, 323
school board in Pennsylvania 90
Science of Mind 45, 129, 139, 362
Science of Mind church 129
scientific community 162, 182, 303

scientific method 8, 93, 114, 316
sea of giant sized dupers 37
The Secret 46, 113
secret secrets 35
self-consciousness 3, 12, 21, 47-8, 72
sensory deprivation 163, 335
sensory deprived mental aberration 164
sexual excitation 248
sexual fantasy 245
Shakespeare 117, 176, 180
Shaman 119-20
Shanghai 73
Sharia 154, 178
silly uber-beliefs 81
Sisyphus 246
Skinner 259
slaves 1, 207, 226
smallpox 103
smoking pot 116
soccer 33
Social Security 109
Society of Skeptics 125
Somalia 152
SPECT image 31
SPECT scan 28, 84, 286, 335
SPECT scans 6, 287
spirit channelers 86
spiritual experience 83, 129
Spirituality 84-5
St. Teresa of Avila 16
Starbucks 111
status quo 34, 76, 227, 267-8
Statute for Religious Freedom 207
Stephen Colbert 233
Steppes-ponies 74
Steven Vogt 96
street smarts 211
suffrage for women 210
suicide bombers 153, 162
super novas 57
Superman 215

supernatural 10, 18, 31, 43, 51, 55, 69, 84, 114, 122, 125, 164, 168, 175, 289
superpower puppets 239-40
Supreme Court 167, 233, 263, 337
survive in the future 108
surviving 185, 255, 352

T

T-shirt 133
tables of inscribed stone 175
Tarot 113
tax money 90
tax supplements 224
telepathic abilities 66
telepathy 66, 219
telescope 71
Temple of the Dead in Greece 49
Terrorism 165, 184, 352, 362
terrorist 31, 238, 268
Thierry Heidmann 102
Thomas Bowdler 180
Thomas Edison 192
Thomas Jefferson 206-7, 211, 223, 262, 350-1
Thoreau 75
threat of 'gods 142
three ball juggling 27
Timothy Leary 18
tinkling bell 172
tolerance and moderation 149-50, 153
Tom Barnett 219
Tom Friedman 226
tooth fairy 50
topographical 24
Torah 170
Toyota 224
trait 3, 292
translation device 115
trillion-neuro-connection 12
TV dish 134

Twain 57, 297
tweener stage 235
Twitter 218

U

uber-belief xii, 47, 65, 77, 80, 82, 97, 114, 127, 140, 153, 155, 158, 164, 221
uber-belief addictions 127
uber-beliefs xii, 5, 8-10, 14, 23, 30, 35, 48, 61, 69-71, 73, 76-8, 81, 83, 99
ultimate purpose 132
unfettered by evidence 160
unhinged 158
UNI-verse 145
unified field theory 139
unitary mental state 164
United Nations 159, 240
United States Supreme Court 233
University of California at Los Angeles 35
Unweaving the Rainbow 40, 360
urban mythology 18
US Marines 176
US preeminence 190
US Presidents 207
USA as a loser 188
Utah 253

V

vengeance 195-201
Victorian England 243
Viet Nam 55, 152
Voltaire 57, 296-7, 299

W

Walk it off 45
walk softly but carry a big stick 153
war game 243
war hero nodule 238
war on Terror 149
war toy superiority 232
The Washington Post 151

Watchman's Rattle 81, 360
water flea 101
weigh a soul 115
Western Democracies 36
William Condon 123
Winston Churchill 233
wishful thinking 46, 80, 133
Wizard of Oz 107
W.M. Keck 28
world citizens 112
world is not flat 162, 183
WWII a good war 234

X

"x" and "y" axis 6

Y

Yankee boy scouts 152
yellow brick road to OZ 187

Z

zealous missionaries 111
"Zener" cards 140
zephyr 170
Zoloft 156

www.ingramcontent.com/pod-product-compliance
Lightning Source LLC
Chambersburg PA
CBHW031816170526
45157CB00001B/80